10 Full Length STAAR Grade 7 Math Practice Tests

The Practice You Need to Ace the STAAR

Grade 7 Math Test

By

Reza Nazari

ISBN: 978-1-63719-358-7

Published by: Effortless Math Education

www.EffortlessMath.com

Welcome to
STAAR Grade 7 Math Prep
2024

Thank you for choosing Effortless Math for your STAAR Grade 7 Math test preparation and congratulations on making the decision to prepare for the STAAR Grade 7 Math test! It's a remarkable move you are taking, one that shouldn't be diminished in any capacity. That's why you need to use every tool possible to ensure you succeed on the test with the highest possible score, and this extensive practice book is one such tool.

This book will help you prepare for (and even ACE) the STAAR Grade 7 Math test. As test day draws nearer, effective preparation becomes increasingly more important. Thankfully, you have this comprehensive practice book to help you get ready for the test. With this book, you can feel confident that you will be more than ready for the STAAR Grade 7 Math test when the time comes.

First and foremost, it is important to note that this book is a practice book and not a prep book. Every test of this "self-guided math practice book" was carefully developed to ensure that you are making the most effective use of your time while preparing for the test. This up-to-date guide reflects the 2024 test guidelines and will put you on the right track to hone your math skills, overcome exam anxiety, and boost your confidence, so that you do your best to succeed on the STAAR Grade 7 Math test.

This practice book will:

☑ Explain the format of the STAAR Grade 7 Math test.

☑ Describe specific test-taking strategies that you can use on the test.

☑ Provide STAAR Grade 7 Math test-taking tips.

☑ Help you identify the areas in which you need to concentrate your study time.

☑ Offer STAAR Grade 7 Math questions and explanations to help you develop the basic math skills.

☑ Give **realistic and full-length practice tests** (featuring new question types) with detailed answers to help you measure your exam readiness and build confidence.

This practice book contains 10 practice tests to help you succeed on the STAAR Grade 7 Math test. You'll get in-depth instructions on every math topic as well as tips and techniques on how to answer each question type. You'll also get plenty of practice questions to boost your test-taking confidence.

In addition, in the following pages you'll find:

➢ **How to Use This Book Effectively** – This section provides you with step-by-step instructions on how to get the most out of this comprehensive practice book.

➢ **How to study for the STAAR Grade 7 Math Test** – A six-step study program has been developed to help you make the best use of this book and prepare for your STAAR Grade 7 Math test. Here you'll find tips and strategies to guide your study program and help you understand STAAR Grade 7 Math and how to ace the test.

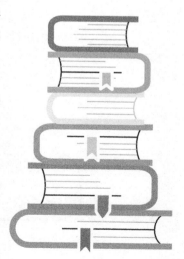

➢ **STAAR Grade 7 Math Review** – Learn everything you need to know about the STAAR Grade 7 Math test.

➢ **STAAR Grade 7 Math Test-Taking Strategies** – Learn how to effectively put these recommended test-taking techniques into use for improving your STAAR Grade 7 Math score.

➢ **Test Day Tips** – Review these tips to make sure you will do your best when the big day comes.

Effortless Math's STAAR Grade 7 Online Center

Effortless Math Online STAAR Grade 7 Center offers a complete study program, including the following:

✓ Step-by-step instructions on how to prepare for the STAAR Grade 7 Math test

✓ Numerous STAAR Grade 7 Math worksheets to help you measure your math skills

✓ Complete list of STAAR Grade 7 Math formulas

✓ Video lessons for all STAAR Grade 7 Math topics

✓ Full-length STAAR Grade 7 Math practice tests

✓ And much more…

No Registration Required.

Visit effortlessmath.com/STAAR7 to find your online STAAR Grade 7 Math resources.

How to Use This Book Effectively

Look no further when you need a practice book to improve your math skills to succeed on the math portion of the STAAR Grade 7 test. Each section of this comprehensive practice book for the STAAR Grade 7 Math will provide you with the knowledge, tools, and understanding needed for every topic covered on the test.

It's important that you understand each topic before moving onto another one, as that's the way to guarantee your success. Each test provides you with sample questions and detail explanation of every concept to better understand the content that will be on the test. To get the best possible results from this book:

➤ **Begin studying long before your test date.** This provides you ample time to learn the different math concepts. The earlier you begin studying for the test, the sharper your skills will be. Do not procrastinate! Provide yourself with plenty of time to learn the concepts and feel comfortable that you understand them when your test date arrives.

➤ **Practice consistently.** Study STAAR Grade 7 Math concepts at least 20 to 30 minutes a day. Remember, slow and steady wins the race, which can be applied to preparing for the STAAR Grade 7 Math test. Instead of cramming to tackle everything at once, be patient and learn the math topics in short bursts.

➤ Whenever you get a math problem wrong, **mark it off, and review it later** to make sure you understand the concept.

➤ Once you've reviewed the book's instructions, **take a practice test** to gauge your level of readiness. Then, review your results. Read detailed answers and solutions for each question you missed.

➤ **Take another practice test** to get an idea of how ready you are to take the actual exam. Taking the practice tests will give you the confidence you need on test day. Simulate the STAAR testing environment by sitting in a quiet room free from distraction.

How to Study for the STAAR Grade 7 Math Test

Studying for the STAAR Grade 7 Math test can be a really daunting and boring task. What's the best way to go about it? Is there a certain study method that works better than others? Well, studying for the STAAR Grade 7 Math can be done effectively. The following six-step program has been designed to make preparing for the STAAR Grade 7 Math test more efficient and less overwhelming.

Step 1 - Create a study plan
Step 2 - Choose your study resources
Step 3 - Review, Learn, Practice
Step 4 - Learn and practice test-taking strategies
Step 5 - Learn the STAAR Grade 7 Test format and take practice tests
Step 6 - Analyze your performance

STEP 1: Create a Study Plan

It's always easier to get things done when you have a plan. Creating a study plan for the STAAR Grade 7 Math test can help you to stay on track with your studies. It's important to sit down and prepare a study plan with what works with your life, classes, and any other obligations you may have. Devote enough time each day to studying. It's also a great idea to break down each section of the exam into blocks and study one concept at a time.

It's important to understand that there is no "right" way to create a study plan. Your study plan will be personalized based on your specific needs and learning style. Follow these guidelines to create an effective study plan for your STAAR Grade 7 Math test:

★ **Analyze your learning style and study habits** – Everyone has a different learning style. It is essential to embrace your individuality and the unique way you learn. Think about what works and what doesn't work for you. Do you prefer STAAR Grade 7 Math prep books or a combination of textbooks and video lessons? Does it work better for you if you study every night for thirty minutes or is it more effective to study in the morning before going to school?

★ **Evaluate your schedule** – Review your current schedule and find out how much time you can consistently devote to STAAR Grade 7 Math study.

★ **Develop a schedule** – Now it's time to add your study schedule to your calendar like any other obligation. Schedule time for study, practice, and review. Plan out which topic you will study on which day to ensure that you're devoting enough time to each concept. Develop a study plan that is mindful, realistic, and flexible.

★ **Stick to your schedule** – A study plan is only effective when it is followed consistently. You should try to develop a study plan that you can follow for the length of your study program.

★ **Evaluate your study plan and adjust as needed** – Sometimes you need to adjust your plan when you have new commitments. Check in with yourself regularly to make sure that you're not falling behind in your study plan. Remember, the most important thing is sticking to your plan. Your study plan is all about helping you be more productive. If you find that your study plan is not as effective as you want, don't get discouraged. It's okay to make changes as you figure out what works best for you.

STEP 2: Choose Your Study Resources

There are numerous textbooks and online resources available for the STAAR Grade 7 Math test, and it may not be clear where to begin. Don't worry! This practice book along with its online resources provides everything you need to fully prepare for your STAAR Grade 7 Math test. In addition to the book content, you can also use Effortless Math's online resources. (video lessons, worksheets, formulas, etc.)

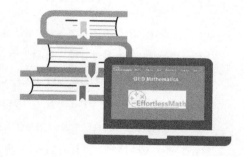

You can visit EffortlessMath.com/STAAR7 to find your online STAAR Grade 7 Math resources.

STEP 3: Review, Learn, Practice

Effortless Math's STAAR course breaks down each subject into specific skills or content areas. For instance, the percent concept is separated into different topics–percent calculation, percent increase and decrease, percent problems, etc. Use our online resources to help you go over all key math concepts and topics on the STAAR Math test.

As you review each concept, take notes or highlight the concepts you would like to go over again in the future. If you're unfamiliar with a topic or something is difficult for you, do additional research on it. For each math topic, plenty of instructions, step-by-step guides, and examples are provided to ensure you get a good grasp of the material. You can also find video lessons on the Effortless Math website for each STAAR Math concept.

Quickly review the topics you do understand to get a brush-up of the material. Be sure to use the worksheets and do the practice questions provided on the Effortless Math's online center to measure your understanding of the concepts.

STEP 4: Learn and Practice Test-taking Strategies

In the following sections, you will find important test-taking strategies and tips that can help you earn extra points. You'll learn how to think strategically and when to guess if you don't know the answer to a question. Using STAAR Grade 7 Math test-taking strategies and tips can help you raise your score and do well on the test. Apply test taking strategies on the practice tests to help you boost your confidence.

STEP 5: Learn the STAAR Grade 7 Test Format and Take Practice Tests

The *STAAR Grade 7 Test Review* section provides information about the structure of the STAAR Grade 7 test. Read this section to learn more about the STAAR test structure, different test sections, the number of questions in each section, and the section time limits. When you have a prior understanding of the test format and different types of STAAR Grade 7 Math questions, you'll feel more confident when you take the actual exam.

Once you have read through the instructions and lessons and feel like you are ready to go – take advantage of all STAAR Grade 7 Math full length practice tests available in this book. Use the practice tests to sharpen your skills and build confidence.

The STAAR Grade 7 Math practice tests are formatted similarly to the actual STAAR Grade 7 Math test. When you take each practice test, try to simulate actual testing conditions. To take the practice tests, sit in a quiet space, time yourself, and work through as many of the questions as time allows. The practice tests are followed by detailed answer explanations to help you find your weak areas, learn from your mistakes, and raise your STAAR Grade 7 Math score.

STEP 6: Analyze Your Performance

After taking each practice test, look over the answer keys and explanations to learn which questions you answered correctly and which you did not. Never be discouraged if you make a few mistakes. See them as a learning opportunity. This will highlight your strengths and weaknesses.

You can use the results to determine if you need additional practice or if you are ready to take the actual STAAR Grade 7 Math test.

Looking for more?

Visit effortlessmath.com/STAAR7 to find hundreds of STAAR Grade 7 Math worksheets, video tutorials, practice tests, STAAR Grade 7 Math formulas, and much more.

No Registration Required.

STAAR Test Review

The State of Texas Assessments of Academic Readiness (STAAR) is a test series supervised by the Texas Education Agency for students in public schools across Texas, from grade 3 to grade 12. These tests track students' progress from grades 3 to 8 and also in high school. STAAR is based on state curriculum standards and assesses core subjects such as:

- Reading
- Writing
- Math
- Science
- Social Studies

The STAAR Grade 7 Math Test consists of multiple-choice and open-ended questions, requiring students to give their own numerical answers. The exam covers a wide range of math topics like number operations, algebra, geometry, measurement, data analysis, and probability. Questions are designed to evaluate students' problem-solving skills, conceptual understanding, and procedural fluency.

The content of the STAAR Grade 7 Math Test is in line with the TEKS curriculum, ensuring the test remains relevant to classroom learning. It covers important math concepts that students should master by the end of grade 7, getting them ready for more advanced math in high school. The variety of topics allows for a thorough assessment of students' math knowledge.

The difficulty level of the STAAR Grade 7 Math Test is well-balanced, providing students with a range of questions that address different skill levels. The test includes a mix of straightforward and complex problems, which lets students demonstrate their understanding of various math concepts. Some real-world problems are also included, promoting critical thinking and practical application of math skills. The test is an effective measure of students' preparedness for high school math and offers valuable insights for teachers to enhance classroom instruction.

Mathematical Process Standards

7.1(A)	7.1(B)	7.1(C)	7.1(D)	7.1(E)	7.1(F)	7.1(G)
apply mathematics to problems arising in everyday life, society, and the workplace	use a problem-solving model that incorporates analyzing given information, formulating a plan or strategy, determining a solution, justifying the solution, and evaluating the problem-solving process and the reasonableness of the solution	select tools, including real objects, manipulatives, paper and pencil, and technology as appropriate, and techniques, including mental math, estimation, and number sense as appropriate, to solve problems	communicate mathematical ideas, reasoning, and their implications using multiple representations, including symbols, diagrams, graphs, and language as appropriate	create and use representations to organize, record, and communicate mathematical ideas	analyze mathematical relationships to connect and communicate mathematical ideas	display, explain, and justify mathematical ideas and arguments using precise mathematical language in written or oral communication

Rptg Cat	STAAR		Readiness Standards		Supporting Standards
1 Probability and Numerical Representations	9	7.6(H)	solve problems using qualitative and quantitative predictions and comparisons from simple experiments.	7.2(A)	extend previous knowledge of sets and subsets using a visual representation to describe relationships between sets of rational numbers.
		7.6(I)	determine experimental and theoretical probabilities related to simple and compound events using data and sample spaces	7.6(A)	represent sample spaces for simple and compound events using lists and tree diagrams.
				7.6(C)	make predictions and determine solutions using experimental data for simple and compound events.
				7.6(D)	make predictions and determine solutions using theoretical probability for simple and compound events.
				7.6(E)	find the probabilities of a simple event and its complement and describe the relationship between the two

2	Computations and Algebraic Relationships	20	7.3(B)	apply and extend previous understandings of operations to solve problems using addition, subtraction, multiplication, and division of rational numbers	7.3(A)	add, subtract, multiply, and divide rational numbers fluently.
			7.4(A)	represent constant rates of change in mathematical and real-world problems given pictorial, tabular, verbal, numeric, graphical, and algebraic representations, including $d = rt$	7.4(B)	calculate unit rates from rates in mathematical and real-world problems.
			7.4(D)	solve problems involving ratios, rates, and percents, including multi-step problems involving percent increase and percent decrease, and financial literacy problems.	7.4(C)	determine the constant of proportionality ($k = y/x$) within mathematical and real-world problems
			7.7(A)	represent linear relationships using verbal descriptions, tables, graphs, and equations that simplify to the form $y = mx + b$	7.10(A)	write one-variable, two-step equations and inequalities to represent constraints or conditions within problems.
			7.11(A)	model and solve one-variable, two-step equations and inequalities	7.10(B)	represent solutions for one-variable, two-step equations and inequalities on number lines.
					7.10(C)	write a corresponding real-world problem given a one-variable, two-step equation or inequality.
					7.11(B)	determine if the given value(s) make(s) one-variable, two-step equations and inequalities true
3	Geometry and Measurement	16	7.5(C)	solve mathematical and real-world problems involving similar shape and scale drawings.	7.4(E)	Convert between measurement systems, including the use of proportions and the use of unit rates.
			7.9(A)	solve problems involving the volume of rectangular prisms, triangular prisms, rectangular pyramids, and triangular pyramids.	7.5(A)	generalize the critical attributes of similarity, including ratios within and between similar shapes.
			7.9(B)	determine the circumference and area of circles.	7.5(B)	describe π as the ratio of the circumference of a circle to its diameter.
			7.9(C)	determine the area of composite figures containing combinations of rectangles, squares, parallelograms, trapezoids, triangles, semicircles, and quarter circles	7.9(D)	solve problems involving the lateral and total surface area of a rectangular prism, rectangular pyramid, triangular prism, and triangular pyramid by determining the area of the shape's net.
					7.11(C)	write and solve equations using geometry concepts, including the sum of the angles in a triangle, and angle relationships

4 Data Analysis and Personal Financial Literacy	9	7.6(G)	solve problems using data represented in bar graphs, dot plots, and circle graphs, including part-to-whole and part-to-part comparisons and equivalents.	7.12(B)	use data from a random sample to make inferences about a population.
		7.12(A)	compare two groups of numeric data u sing comparative dot plots or box plots by comparing their shapes, centers, and spreads	7.12(C)	compare two populations based on data in random samples from these populations, including informal comparative inferences about differences between the two populations.
				7.13(A)	calculate the sales tax for a given purchase and calculate income tax for earned wages.
				7.13(B)	identify the components of a personal budget, including income, planned savings for college, retirement, and emergencies; taxes; and fixed and variable expenses, and calculate what percentage each category comprises of the total budget.
				7.13(C)	create and organize a financial assets and liabilities record and construct a net worth statement.
				7.13(D)	use a family budget estimator to determine the minimum household budget and average hourly wage needed for a family to meet its basic needs in the student's city or another large city nearby.
				7.13(E)	calculate and compare simple interest and compound interest earnings.
				7.13(F)	analyze and compare monetary incentives, including sales, rebates, and coupons
# Items	54 (4 Gradable)	32-35 questions from Readiness Standards		19-22 questions from Supporting Standards	

STAAR Grade 7 Math Test-Taking Strategies

Here are some test-taking strategies that you can use to maximize your performance and results on the STAAR Grade 7 Math test.

#1: Use This Approach To Answer Every STAAR Grade 7 Math Question

- Review the question to identify keywords and important information.

- Translate the keywords into math operations so you can solve the problem.

- Review the answer choices. What are the differences between answer choices?

- Draw or label a diagram if needed.

- Try to find patterns.

- Find the right method to answer the question. Use straightforward math, plug in numbers, or test the answer choices (backsolving).

- Double-check your work.

#2: Use Educated Guessing

This approach is applicable to the problems you understand to some degree but cannot solve using straightforward math. In such cases, try to filter out as many answer choices as possible before picking an answer. In cases where you don't have a clue about what a certain problem entails, don't waste any time trying to eliminate answer choices. Just choose one randomly before moving onto the next question.

As you can ascertain, direct solutions are the most optimal approach. Carefully read through the question, determine what the solution is using the math you have learned before, then coordinate the answer with one of the choices available to you. Are you stumped? Make your best guess, then move on.

Don't leave any fields empty! Even if you're unable to work out a problem, strive to answer it. Take a guess if you have to. You will not lose points by getting an answer wrong, though you may gain a point by getting it correct!

#3 : BALLPARK

A ballpark answer is a rough approximation. When we become overwhelmed by calculations and figures, we end up making silly mistakes. A decimal that is moved by one unit can change an answer from right to wrong, regardless of the number of steps that you went through to get it. That's where ballparking can play a big part.

If you think you know what the correct answer may be (even if it's just a ballpark answer), you'll usually have the ability to eliminate a couple of choices. While answer choices are usually based on the average student error and/or values that are closely tied, you will still be able to weed out choices that are way far afield. Try to find answers that aren't in the proverbial ballpark when you're looking for a wrong answer on a multiple-choice question. This is an optimal approach to eliminating answers to a problem.

#4 : BACKSOLVING

A majority of questions on the STAAR Grade 7 Math test will be in multiple-choice format. Many test-takers prefer multiple-choice questions, as at least the answer is right there. You'll typically have four answers to pick from. You simply need to figure out which one is correct. Usually, the best way to go about doing so is "backsolving."

As mentioned earlier, direct solutions are the most optimal approach to answering a question. Carefully read through a problem, calculate a solution, then correspond the answer with one of the choices displayed in front of you. If you can't calculate a solution, your next best approach involves "backsolving."

When backsolving a problem, contrast one of your answer options against the problem you are asked, then see which of them is most relevant. More often than not, answer choices are listed in ascending or descending order. In such cases, try out the choices B or C. If it's not correct, you can go either down or up from there.

#5 : PLUGGING IN NUMBERS

"Plugging in numbers" is a strategy that can be applied to a wide range of different math problems on the STAAR Grade 7 Math test. This approach is typically used to simplify a challenging question so that it is more understandable. By using the strategy carefully, you can find the answer without too much trouble.

The concept is fairly straightforward–replace unknown variables in a problem with certain values. When selecting a number, consider the following:

- Choose a number that's basic (just not too basic). Generally, you should avoid choosing 1 (or even 0). A decent choice is 2.

- Try not to choose a number that is displayed in the problem.

- Make sure you keep your numbers different if you need to choose at least two of them.

- More often than not, choosing numbers merely lets you filter out some of your answer choices. As such, don't just go with the first choice that gives you the right answer.

- If several answers seem correct, then you'll need to choose another value and try again. This time, though, you'll just need to check choices that haven't been eliminated yet.

- If your question contains fractions, then a potential right answer may involve either an LCD (least common denominator) or an LCD multiple.

- 100 is the number you should choose when you are dealing with problems involving percentages.

STAAR Grade 7 Math – Test Day Tips

After practicing and reviewing all the math concepts you've been taught, and taking some STAAR Grade 7 mathematics practice tests, you'll be prepared for test day. Consider the following tips to be extra-ready come test time.

Before Your Test

What to do the night before:

- **Relax!** One day before your test, study lightly or skip studying altogether. You shouldn't attempt to learn something new, either. There are plenty of reasons why studying the evening before a big test can work against you. Put it this way–a marathoner wouldn't go out for a sprint before the day of a big race. Mental marathoners–such as yourself–should not study for any more than one hour 24 hours before a STAAR Grade 7 test. That's because your brain requires some rest to be at its best. The night before your exam, spend some time with family or friends, or read a book.

- **Avoid bright screens** - You'll have to get some good shuteye the night before your test. Bright screens (such as the ones coming from your laptop, TV, or mobile device) should be avoided altogether. Staring at such a screen will keep your brain up, making it hard to drift asleep at a reasonable hour.

- **Make sure your dinner is healthy** - The meal that you have for dinner should be nutritious. Be sure to drink plenty of water as well. Load up on your complex carbohydrates, much like a marathon runner would do. Pasta, rice, and potatoes are ideal options here, as are vegetables and protein sources.

- **Get your bag ready for test day** - The night prior to your test, pack your bag with your stationery, admissions pass, ID, and any other gear that you need. Keep the bag right by your front door.

- **Make plans to reach the testing site** - Before going to sleep, ensure that you understand precisely how you will arrive at the site of the test. If parking is something you'll have to find first, plan for it. If you're dependent on public transit, then review the schedule. You should also make sure that the train/bus/subway/streetcar you use will be running. Find out about road closures as well. If a parent or friend is accompanying you, ensure that they understand what steps they have to take as well.

The Day of the Test

- **Get up reasonably early, but not too early.**

- **Have breakfast** - Breakfast improves your concentration, memory, and mood. As such, make sure the breakfast that you eat in the morning is healthy. The last thing you want to be is distracted by a grumbling tummy. If it's not your own stomach making those noises, another test taker close to you might be instead. Prevent discomfort or embarrassment by consuming a healthy breakfast. Bring a snack with you if you think you'll need it.

- **Follow your daily routine** - Do you watch Good Morning America each morning while getting ready for the day? Don't break your usual habits on the day of the test. Likewise, if coffee isn't something you drink in the morning, then don't take up the habit hours before your test. Routine consistency lets you concentrate on the main objective–doing the best you can on your test.

- **Wear layers** - Dress yourself up in comfortable layers. You should be ready for any kind of internal temperature. If it gets too warm during the test, take a layer off.

- **Get there on time** - The last thing you want to do is get to the test site late. Rather, you should be there 45 minutes prior to the start of the test. Upon your arrival, try not to hang out with anybody who is nervous. Any anxious energy they exhibit shouldn't influence you.

- **Leave the books at home** - No books should be brought to the test site. If you start developing anxiety before the test, books could encourage you to do some last-minute studying, which will only hinder you. Keep the books far away–better yet, leave them at home.

- **Make your voice heard** - If something is off, speak to a proctor. If medical attention is needed or if you'll require anything, consult the proctor prior to the start of the test. Any doubts you have should be clarified. You should be entering the test site with a state of mind that is completely clear.

■ **Have faith in yourself** - When you feel confident, you will be able to perform at your best. When you are waiting for the test to begin, envision yourself receiving an outstanding result. Try to see yourself as someone who knows all the answers, no matter what the questions are. A lot of athletes tend to use this technique—particularly before a big competition. Your expectations will be reflected by your performance.

During your test

■ **Be calm and breathe deeply** - You need to relax before the test, and some deep breathing will go a long way to help you do that. Be confident and calm. You got this. Everybody feels a little stressed out just before an evaluation of any kind is set to begin. Learn some effective breathing exercises. Spend a minute meditating before the test starts. Filter out any negative thoughts you have. Exhibit confidence when having such thoughts.

■ **Concentrate on the test** - Refrain from comparing yourself to anyone else. You shouldn't be distracted by the people near you or random noise. Concentrate exclusively on the test. If you find yourself irritated by surrounding noises, earplugs can be used to block sounds off close to you. Don't forget—the test is going to last several hours if you're taking more than one subject of the test. Some of that time will be dedicated to brief sections. Concentrate on the specific section you are working on during a particular moment. Do not let your mind wander off to upcoming or previous sections.

■ **Skip challenging questions** - Optimize your time when taking the test. Lingering on a single question for too long will work against you. If you don't know what the answer is to a certain question, use your best guess, and mark the question so you can review it later on. There is no need to spend time attempting to solve something you aren't sure about. That time would be better served handling the questions you can actually answer well. You will not be penalized for getting the wrong answer on a test like this.

■ **Try to answer each question individually** - Focus only on the question you are working on. Use one of the test-taking strategies to solve the problem. If you aren't able to come up with an answer, don't get frustrated. Simply skip that question, then move onto the next one.

- **Don't forget to breathe!** Whenever you notice your mind wandering, your stress levels boosting, or frustration brewing, take a thirty-second break. Shut your eyes, drop your pencil, breathe deeply, and let your shoulders relax. You will end up being more productive when you allow yourself to relax for a moment.

- **Review your answer.** If you still have time at the end of the test, don't waste it. Go back and check over your answers. It is worth going through the test from start to finish to ensure that you didn't make a sloppy mistake somewhere.

- **Optimize your breaks** - When break time comes, use the restroom, have a snack, and reactivate your energy for the subsequent section. Doing some stretches can help stimulate your blood flow.

After your test

- **Take it easy** - You will need to set some time aside to relax and decompress once the test has concluded. There is no need to stress yourself out about what you could've said, or what you may have done wrong. At this point, there's nothing you can do about it. Your energy and time would be better spent on something that will bring you happiness for the remainder of your day.

Contents

STAAR Mathematics

Practice Test 1

2024

Grade 7

Total number of questions: 40

Total time to complete the test: No time limit

You may NOT use a calculator.

1

STAAR Grade 7 Mathematics Formula Sheet

LINEAR EQUATIONS

Slope – intercept form

$$y = mx + b$$

Direct Variation

$$y = kx$$

Slope of a Line

$$m = \frac{y_2 - y_1}{x_2 - x_1}$$

CIRCUMFERENCE

Circle

$$C = 2\pi r \text{ or } C = \pi d$$

AREA

Triangle

$$A = \frac{1}{2}bh$$

Parallelogram

$$A = bh$$

Trapezoid

$$A = \frac{1}{2}h(b_1 + b_2)$$

Circle

$$A = \pi r^2$$

SURFACE AREA

	Lateral	Total
Prism	$S = Ph$	$S = Ph + 2B$
Cylinder	$S = 2\pi rh$	$S = 2\pi rh + 2\pi r^2$

VOLUME

Prism or Cylinder

$$V = Bh$$

Pyramid or Cone

$$V = \frac{1}{3}Bh$$

Sphere

$$V = \frac{4}{3}\pi r^3$$

ADDITIONAL INFORMATION

Pythagorean theorem

$$a^2 + b^2 = c^2$$

Simple interest

$$I = prt$$

Compound Interest

$$A = p(1 + r)^t$$

1) The table below shows the number of students who participated in a survey about their favorite school subjects. A student will be randomly selected from the survey participants.

Subject	Number of Students
Math	30
Science	20
English	40
History	10

Which statement about the favorite subject of the randomly selected student is best supported by the information in the table?

A. The favorite subject is most likely to be English.

B. The favorite subject is twice as likely to be math as it is to be science.

C. The favorite subject is equally likely to be math, science, English, or history.

D. The favorite subject is more than twice as likely to be history as it is to be science.

2) The length of a rope is 15 feet. There are approximately 0.3048 meters in 1 foot. Which measurement is closest to the length of the rope in meters?

A. $4.57\ m$

B. $45.7\ m$

C. $457\ m$

D. $0.457\ m$

3) Angle A and Angle B are complementary angles.

The measure of angle A is $35°$.

The measure of angle B is $(6y - 11)°$.

Which equation can be used to find the value of y?

A. $35 + (6y - 11) = 90$

B. $35 = 6y - 11$

C. $35 + (6y - 11) = 180$

D. $35 + (6y - 11) = 360$

4) The price of a book is $13. The sales tax rate is 6%. What is the sales tax on this book in dollars and cents?

A. 7.8

B. 0.78

C. 12.22

D. 5.2

5) A car traveled 84 miles in 6 hours. At this rate, how many miles will the car travel in $\frac{1}{3}$ hour?

A. 4 miles

B. $3\frac{1}{3}$ miles

C. $4\frac{2}{3}$ miles

D. 5 miles

6) The figure below consists of a triangle and a trapezoid. What is the area of the figure in square meters?

A. 120

B. 271

C. 200

D. 211

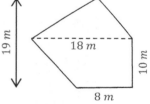

7) Which number line represents the solution to $5x + 3 \geq -2$?

A.

B.
```
←————————○——————————————→
 -6  -5  -4  -3  -2  -1   0   1   2   3   4   5   6
```

C.
```
←————————○————————————————→
 -6  -5  -4  -3  -2  -1   0   1   2   3   4   5   6
```

D.
```
←—————————————————●————————→
 -6  -5  -4  -3  -2  -1   0   1   2   3   4   5   6
```

8) A jar contains 10 red, 7 blue, and 5 yellow marbles. If you select two marbles at random without replacement, what is the probability that both marbles will be blue?

A. $\dfrac{1}{17}$

B. $\dfrac{7}{90}$

C. $\dfrac{7}{51}$

D. $\dfrac{1}{11}$

9) Find the value of x according to the following model.

A. $x = \dfrac{5}{6}$

B. $x = \dfrac{1}{6}$

C. $x = \dfrac{2}{6}$

D. $x = \dfrac{3}{6}$

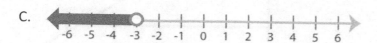

10) What is the result of the following expression: $-12.75 \div \left(3\dfrac{3}{4}\right)$?

11) The following triangles are similar. Which proportion can be used to calculate the length of the side DE?

A. $\dfrac{17}{8.5} = \dfrac{15}{DE}$

B. $\dfrac{25.5}{8} = \dfrac{15}{DE}$

C. $\dfrac{8}{12} = \dfrac{15}{DE}$

D. $\dfrac{15}{17} = \dfrac{DE}{12}$

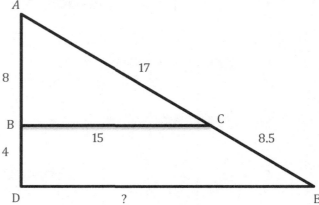

12) A store has a total budget of $1,000 to buy some chairs and tables for their new restaurant. The chairs cost $50 each and the tables cost $200 each. Let c represent the number of chairs and t represent the number of tables. Which inequality represents all possible values of t and c, the number of tables and chairs they can buy with the given budget?

A. $200\,t + 50\,c \le 1,000$

B. $200\,t + 50\,c \ge 1,000$

C. $200\,t + 50\,c < 1,000$

D. $200\,t + 50\,c > 1,000$

13) What is the equation of the line shown in the graph below?

A. $y = 3x + 2$

B. $y = -2x + 3$

C. $y = 2x + 3$

D. $y = -2x - 3$

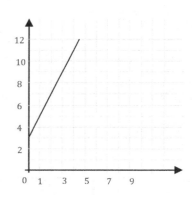

14) The net of a rectangular prism is shown in the diagram. What is its surface area?

A. 98 cm^2

B. 140 cm^2

C. 108 cm^2

D. 216 cm^2

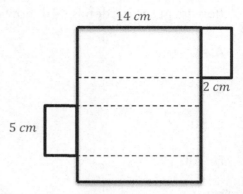

14 cm

2 cm

5 cm

15) The following net worth statement shows Jane's financial situation. Assets are represented with positive values, and liabilities are shown with negative values. However, the current value of her car is not given.

Item	Value
Car (current value)	
Savings account	$3,500
Credit-card debt	−$1,200
Investments	$10,000
Mortgage	−$120,000
401(k) plan	$45,000
Personal loan	−$5,000

If Jane's net worth is $12,300, what is the current value of Jane's car?

A. $67,700

B. $81,000

C. $80,000

D. $83,200

16) Mangoes are sold at $2.50 per piece. Which equation best represents y, the total cost of x pieces of mangoes?

A. $x = 2.50y$

B. $x = 2.50 + y$

C. $y = 2.50 + x$

D. $y = 2.50x$

17) In a garden, the ratio of red to white flowers is 2 to 5. If the total number of red and white flowers in the garden is 140, how many red flowers are there in the garden?

A. 40

B. 20

C. 50

D. 100

18) 120 people were asked about their favorite color. The graph below shows the interest percentage of these people in different colors. How many more people like blue than red?

A. 18

B. 30

C. 12

D. 22

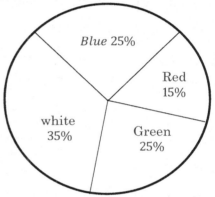

19) Mandy walked for 1 hour at a speed of $4\frac{km}{h}$, then ran for 2 hours at a speed of $8\frac{km}{h}$, and then walked again for 1 hour at a speed of $3\frac{km}{h}$. What was the total distance Mandy traveled during this time?

A) 20 km

B) 23 km

C) 25 km

D) 28 km

20) What is the volume of the following figure?

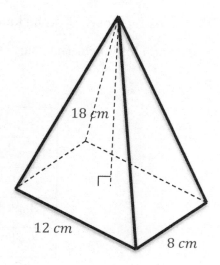

A. 576 cm^3

B. 675 cm^3

C. 887 cm^3

D. 1,728 cm^3

21) According to the figure, what is the value of x?

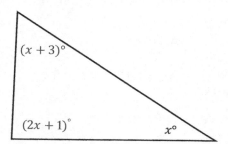

22) The original price of a jacket is $80. If the price is increased to $100, what is the percentage increase?

A. 20%

B. 25%

C. 50%

D. 60%

23) In the figure below, the radius of the small circle is 5 cm and the diameter of the big circle is 20 cm. What is the area of the shaded part?

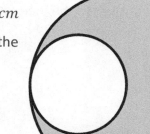

A. 25π cm^2

B. 100π cm^2

C. 75π cm^2

D. 125π cm^2

24) The table shows the prices of some lunch items at a café. John ordered a sandwich, a bag of chips, and a soda for lunch. The sales tax for the order was $0.75. He paid for his lunch with a $20 bill.

Lunch Menu	
Item	Price
Sandwich	$6.99
Bag of chips	$1.49
Soda	$1.89

How much change should John receive from the $20 bill?

A. $8.88

B. $11.12

C. $11.76

D. $12.35

25) Which equation is true when $x = 2$?

A. $2x + 4 = 8$

B. $4x - 6 = 10$

C. $3x + 7 = 16$

D. $5x + 3 = 14$

26) What is the answer to the inequality $2x + 5 > 17$?

A. $x < 6$

B. $x > 6$

C. $x > 3$

D. $x < 3$

27) Which expression represents an arithmetic sequence where the common difference is 5 and the first term is -2?

   A. $-2, 5, 10, 15, 20, \ldots$

   B. $-2, 0, 5, 10, 15, \ldots$

   C. $-2, -7, -12, -17, -22, \ldots$

   D. $-2, 3, 8, 13, 18, \ldots$

28) The temperature in a city was recorded for five consecutive days: $20°C, 21°C, 19°C, 18°C, 22°C$. Which measure of data would best describe the variability of these temperature readings?

   A. Mean

   B. Median

   C. Mode

   D. Range

29) What is the volume of the following triangular prism?

   A. $54 \ m^3$

   B. $28 \ m^3$

   C. $70 \ m^3$

   D. $140 \ m^3$

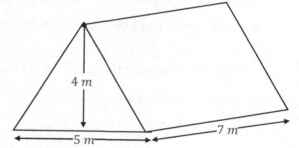

30) An artist created a painting of a landscape using a scale in which 1 inch represents 10 miles. The actual distance between two mountains in the landscape is 70 miles. What is the distance between the two mountains in the painting, in inches?

   A. 0.7 inches.

   B. 7 inches.

   C. 700 inches.

   D. 10 inches.

31) The box plots below show the number of hours students spent studying for a test in two different study groups. Which statement is best supported by the information in the box plots?

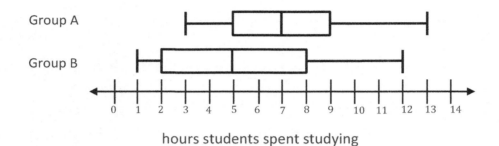

hours students spent studying

A. The interquartile range of the data for Study Group A is equal to the interquartile range of the data for Study Group B.

B. The interquartile range of the data for Study Group A is greater than the interquartile range of the data for Study Group B.

C. The median number of hours spent studying in Study Group A is equal to the median number of hours spent studying in Study Group B.

D. The median number of hours spent studying in Study Group A is greater than the median number of hours spent studying in Study Group B.

32) What is the equation that represents the linear relationship between the x −values and the y −values in the given table?

A. $y = 2x$

B. $y = x + 2$

C. $y = 3x - 1$

D. $y = 2x + 2$

x	y
2	6
3	8
5	12
6	14

33) The average of six numbers is 25. If a seventh number, 46, is added, then, what is the new average?

34) Which of the following is equivalent to $1 < -2x + 5 < 15$

A. $-7 < x < 0$

B. $0 < x < 12$

C. $-5 < x < 2$

D. $-4 < x < 10$

35) According to the following figure, which point is inside the triangle?

A. $(2, -1)$

B. $(0, 1)$

C. $(-2, -1)$

D. $(3, -4)$

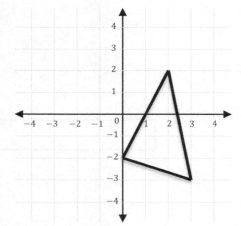

36) Every week, a soccer team practices for 4 hours on Monday, 3 hours on Wednesday, and 2 hours on Friday. At this rate, how many total hours will the team practice in 3 weeks?

A. 18 hours

B. 27 hours

C. 36 hours

D. 45 hours

37) The figure below consists of a triangle, a rectangle, and two semicircles. What is the total area of the shape?

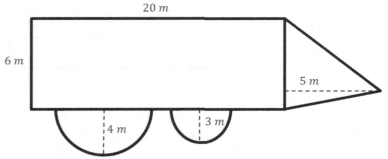

A. $192.56 \ m^2$

B. $168.32 \ m^2$

C. $120 \ m^2$

D. $174.25 \ m^2$

38) What is the median of these numbers? 5,12,8,21,27,14,24,0,33

A. 14

B. 21

C. 17.5

D. 13

39) The cube root of 1,728 is?

40) When a number is subtracted from 35, and the difference is divided by that number, the result is 4. What is the value of the number?

A. 7

B. 6

C. 5

D. 8

End of STAAR Grade 7 Math Practice Test 1

STAAR Mathematics

Practice Test 2

2024

Grade 7

Total number of questions: 40

Total time to complete the test: No time limit

<u>**You may NOT use a calculator.**</u>

15

STAAR Grade 7 Mathematics Formula Sheet

LINEAR EQUATIONS

Slope – intercept form

$$y = mx + b$$

Direct Variation

$$y = kx$$

Slope of a Line

$$m = \frac{y_2 - y_1}{x_2 - x_1}$$

CIRCUMFERENCE

Circle

$$C = 2\pi r \text{ or } C = \pi d$$

AREA

Triangle

$$A = \frac{1}{2} bh$$

Parallelogram

$$A = bh$$

Trapezoid

$$A = \frac{1}{2} h(b_1 + b_2)$$

Circle

$$A = \pi r^2$$

SURFACE AREA

	Lateral	Total
Prism	$S = Ph$	$S = Ph + 2B$
Cylinder	$S = 2\pi rh$	$S = 2\pi rh + 2\pi r^2$

VOLUME

Prism or Cylinder

$$V = Bh$$

Pyramid or Cone

$$V = \frac{1}{3} Bh$$

Sphere

$$V = \frac{4}{3} \pi r^3$$

ADDITIONAL INFORMATION

Pythagorean theorem

$$a^2 + b^2 = c^2$$

Simple interest

$$I = prt$$

Compound Interest

$$A = p(1 + r)^t$$

1) What is the volume of the following cylinder?

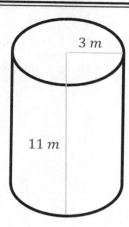

A. $207.24\ m^3$

B. $310.86\ m^3$

C. $1,139.82\ m^3$

D. $103.62\ m^3$

2) A bag contains 20 balls, 5 of which are black, 8 that are red, and 7 that are white. If three balls are drawn at random without replacement, what is the probability that all three balls will be red?

A. $\dfrac{14}{285}$

B. $\dfrac{14}{323}$

C. $\dfrac{56}{969}$

D. $\dfrac{7}{66}$

3) The figure below consists of a triangle, a rectangle, and two trapezoids. What is the total area of the shape?

A. $147.5\ m^2$

B. $165\ m^2$

C. $182\ m^2$

D. $199.5\ m^2$

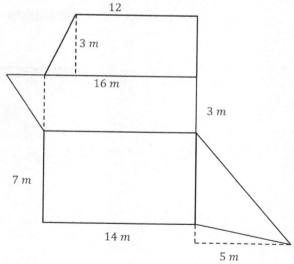

4) A square pyramid has a height of 10 inches, and each side of the base is 8 inches. What is the volume of the pyramid measured in cubic inches?

A) $160 \ in^3$

B) $213.33 \ in^3$

C) $320 \ in^3$

D) $426.67 \ in^3$

5) What is the value of y in the equation $\frac{3}{4}y - \frac{1}{8} = \frac{5}{16}$?

6) Which of the following graphs represents the compound inequality $-5 \le 3x - 2 < 4$?

A.

B.

C.

D.

7) Based on the following information, what is the net worth of John?

Net Worth Statement	
Assets	**Value**
Savings account	$10,000
Investments	$35,000
House (current value)	$200,000
Car	$15,000
Total Assets	**$260,000**
Liabilities	
Mortgage	$100,000
Car loan	$7,000
Credit card debt	$3,500
Total Liabilities	**$110,500**

A. $260,000

B. $149,500

C. $110,500

D. $149,500

8) The following table represents the value of x and the corresponding function $f(x)$. Which of the following could be the equation of the function $f(x)$?

A. $f(x) = 4x^2 + 1$

B. $f(x) = x^2 + 1$

C. $f(x) = \sqrt{x} + 3$

D. $f(x) = 3x^2 + 1$

x	$f(x)$
1	4
2	13
3	28
4	49

9) Triangle ABC and triangle EFG are similar. Which proportion can be used to find the length of FG?

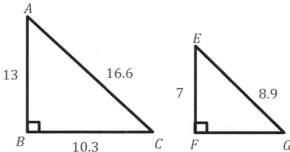

A. $\dfrac{13}{10.3} = \dfrac{7}{FG}$

B. $\dfrac{13}{16.6} = \dfrac{FG}{7}$

C. $\dfrac{16.6}{13} = \dfrac{7}{FG}$

D. $\dfrac{13}{10.3} = \dfrac{8.9}{FG}$

10) Which values from the set $\{1, -2, 0, 3, -4, 10\}$ satisfy this inequality? $4x - 3 > 2$

A. $-2, -4, 0$

B. 10 only

C. 1,3,10,0

D. 3,10

11) What is the surface area of the figure below?

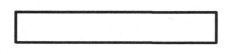

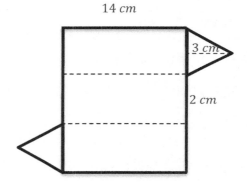

12) Two stores are having sales on shoes.

• At Store P, all shoes are on sale for 30% off the original price.

• At Store Q, all shoes are on sale for a flat discount of $50.

Which store would offer a better sale price for a pair of shoes with an original price of $150?

A. Store P, because the sale price would be $105.

B. Store Q, because the sale price would be $100.

C. Store P, because the sale price would be $50.

D. Store Q, because the sale price would be $105.

13) Which arithmetic sequence is represented by the expression $4x + 6$, where x represents the position of a term in the sequence?

    A. 10,13,16,19,22, …

    B. 10,14,18,22, …

    C. 0,6,12,18,24, …

    D. 4,6,10,14,18, …

14) Given the path shown in the coordinate plane below, which ordered pair represents a point on the path?

    A. $(4, 4)$

    B. $(1, 3)$

    C. $(1, -3)$

    D. $(6, -1)$

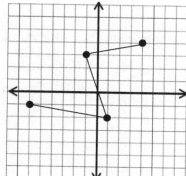

15) In a chemical solution that contains 15% salt, there are 60 g of salt. What is the weight of the solution?

16) A rectangular room measures 10 feet by 12 feet. Given that approximately 0.3048 meters equate to 1 foot, which measurement is closest to the area of the room in square meters?

    A. $11.15 \ m^2$

    B. $120 \ m^2$

    C. $13.5 \ m^2$

    D. $9.65 \ m^2$

17) A survey was conducted among 250 students to determine their preferred subject in school. The following chart shows the percentage of students who preferred each subject. How many more students prefer Math than History?

A. 30

B. 50

C. 65

D. 40

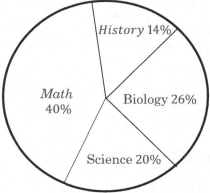

18) The table below displays the number of hours per week that a group of students spend studying. What is the difference between the mean and mode of this data?

A. 0.29

B. 0.43

C. 3

D. 1.2

Student	Hours
A	2
B	3
C	4
D	4
E	5
F	6
G	7

19) The square of a number is $\frac{16}{49}$. What is the cube of that number?

20) What is the value of the expression $5(2x + 4y) + (3 + x)^2$ when $x = 3$ and $y = 2$?

A. 70

B. 34

C. 106

D. 55

21) What is the result of the following expression: $-16.9 \div \left(4\frac{1}{3}\right)$?

    A. -4

    B. -3.9

    C. 3.87

    D. 4

22) Figure ABC and figure $A'B'C'$ are shown on the coordinate grid below. Which statement describes how figure ABC was transformed to form the image $A'B'C'$?

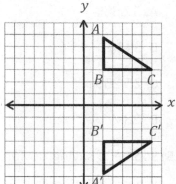

    A. A reflection across the x −axis

    B. A reflection across the origin

    C. A translation across the x −axis

    D. A reflection across the y −axis

23) A company has a budget of $5,000 to buy some computers and printers. The computers cost $1,000 each and the printers cost $500 each. Let c represent the number of computers and p represent the number of printers. Which inequality represents all possible values of c and p, the number of computers and printers they can buy with the given budget?

    A. $1000\,c + 500\,p \le 5,000$

    B. $1000\,c + 500\,p \ge 5,000$

    C. $1000\,c + 500\,p < 5,000$

    D. $1000\,c + 500\,p > 5,000$

24) Mariah has a monthly income of $6,000 and her monthly expenses are categorized as follows:

Monthly Budget	
Category	Amount of Money
Rent	$1,800
Groceries	$450
Savings	$1,200
Car payment	$900
Utilities	$180

What percentage of Mariah's income is used to pay for her rent and groceries?

A. 37.5%

B. 31%

C. 25.25%

D. 23%

25) Sophie is saving up for a new laptop. She starts with $500 and saves $100 each month. Which equation can be used to find y, the total amount of money Sophie will have saved after x months?

A. $y = 100x + 500$

B. $y = 500x + 100$

C. $y = 100x - 500$

D. $y = 500 - 100x$

26) Sophie bought 12 boxes of chocolates for a party.

• Each box of chocolate costs $5.50.

• She paid $10.25 for wrapping the boxes.

What is the total amount Sophie paid?

A. $70.25

B. $66.00

C. $76.25

D. $68.75

27) What is the value of x in the following figure? (Figure not drawn to scale)

A. 15

B. 22

C. 18

D. 41

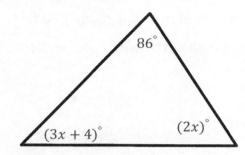

28) Which equation represents the relationship between x and y in the graph?

A. $y = \frac{1}{3}x + 3$

B. $y = -\frac{2}{3}x - 1$

C. $y = -\frac{2}{3}x + 3$

D. $y = \frac{2}{3}x + 1$

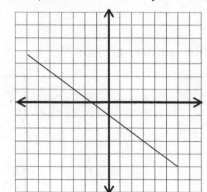

29) In a box of marbles, the ratio of blue to green marbles is 3 to 4. If there are a total of 210 blue and green marbles in the box, how many blue marbles are in the box?

A. 95

B. 75

C. 90

D. 105

30) A restaurant bill comes to $60. The sales tax rate is 7.5%. What is the sales tax on this bill in dollars and cents?

A. 0.45

B. 4.50

C. 7.05

D. 3.80

31) An architect created a blueprint of a house using a scale in which 1 inch represents 1 foot.
The actual length of the living room is 20 feet.
What is the length of the living room in the blueprint, in inches?

A. 2 inches.

B. 10 inches.

C. 20 inches.

D. 200 inches.

32) What is the area of the shaded region?

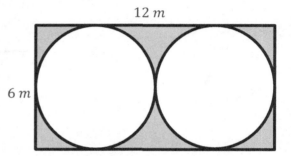

A. $72 - 9\pi \ m^2$

B. $72 - 18\pi \ m^2$

C. $36 - 18\pi \ m^2$

D. $36 - 9\pi \ m^2$

33) The measure of the angles of a triangle are in the ratio $2:3:7$. What is the measure of the smallest angle?

A. $105°$

B. $45°$

C. $55°$

D. $30°$

34) Tom covered 112 miles in 8 hours, while Sarah covered 80 miles in 5 hours. What is the ratio of the average speed of Tom to the average speed of Sarah?

A. $3:4$

B. $4:3$

C. $7:8$

D. $8:7$

35) What is the answer to the inequality $7x - 12 \leq -8$?

A. $x \leq \frac{4}{7}$

B. $x \geq \frac{4}{7}$

C. $x \leq \frac{4}{5}$

D. $x \leq \frac{3}{4}$

36) Which expression is represented by the model below?

A. $5.(2)$

B. $5.(-2)$

C. $10.(1)$

D. $(2).(4)$

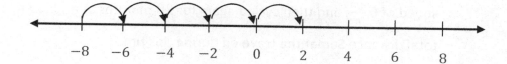

37) If $3x + y = 12$, $6x + 5y = -3$, which of the following ordered pairs (x, y) satisfies both equations?

A. $(7, 7)$

B. $(7, -9)$

C. $(3, 5)$

D. $(1, 2)$

38) A recipe calls for 4 cups of sugar to make 24 servings of a dessert. What is the constant of proportionality that relates the number of servings, y, to the number of cups of sugar used, x?

A. 0.5

B. 4

C. 6

D. 24

39) Jenny has downloaded 240 songs on her phone, which takes up 720 megabytes of space. If each song takes up the same amount of space, how many megabytes would 30 songs take up?

A. 90 MB

B. 72 MB

C. 24 MB

D. 30 MB

40) Samantha cycled for 30 minutes at a speed of $12\frac{km}{h}$, then jogged for 45 minutes at a speed of $6\frac{km}{h}$, and then cycled again for 20 minutes at a speed of $10\frac{km}{h}$. What was the total distance Samantha traveled during this time?

A. $11\ km$

B. $14.3\ km$

C. $9.3\ km$

D. $13.8\ km$

End of STAAR Grade 7 Math Practice Test 2

STAAR Mathematics

Practice Test 3

2024

Grade 7

Total number of questions: 40

Total time to complete the test: No time limit

You may NOT use a calculator.

29

STAAR Grade 7 Mathematics Formula Sheet

LINEAR EQUATIONS	
Slope – intercept form	$y = mx + b$
Direct Variation	$y = kx$
Slope of a Line	$m = \dfrac{y_2 - y_1}{x_2 - x_1}$

CIRCUMFERENCE	
Circle	$C = 2\pi r \text{ or } C = \pi d$

AREA	
Triangle	$A = \dfrac{1}{2}bh$
Parallelogram	$A = bh$
Trapezoid	$A = \dfrac{1}{2}h(b_1 + b_2)$
Circle	$A = \pi r^2$

SURFACE AREA	Lateral	Total
Prism	$S = Ph$	$S = Ph + 2B$
Cylinder	$S = 2\pi rh$	$S = 2\pi rh + 2\pi r^2$

VOLUME	
Prism or Cylinder	$V = Bh$
Pyramid or Cone	$V = \dfrac{1}{3}Bh$
Sphere	$V = \dfrac{4}{3}\pi r^3$

ADDITIONAL INFORMATION	
Pythagorean theorem	$a^2 + b^2 = c^2$
Simple interest	$I = prt$
Compound Interest	$A = p(1 + r)^t$

1) What are the dimensions, in centimeters, of a triangle that is similar to triangle QRS, whose dimensions are given in the diagram below?

A. $6\ cm, 10.5\ cm, 13.5cm$

B. $7\ cm, 14\ cm, 17.5\ cm$

C. $4\ cm, 7.5\ cm, 8.5\ cm$

D. $9\ cm, 18\ cm, 27cm$

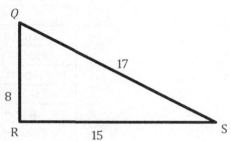

2) Nathan compared the weight of cereal to the amount of sugar per serving of four different brands of cereal. The information he gathered is shown in the table.

Cereal Comparison		
Brand	Weight (oz)	Sugar per Serving (g)
A	12	6
B	16	16
C	10	8
D	14	10

Based on the data in the table, which brand of cereal has the highest amount of sugar per ounce?

A. Brand A

B. Brand B

C. Brand C

D. Brand D

3) Karen drives a taxi and charges $20 for a ride plus an additional $2 per mile driven. Which equation can be used to find y, the total cost of a ride that is x miles long?

A. $y = 20x + 2$

B. $y = 2x + 20$

C. $y = 20x - 2$

D. $y = 2x - 20$

4) Samantha earns a monthly income of $4,500. She allocates her income to different categories as shown in the table below:

Monthly Budget	
Category	Amount of Money
Housing	$1,500
Transportation	$600
Utilities	$600
Food	$750
Savings	$1050

What is the percentage of Samantha's income that goes towards housing and food?

A. 44%

B. 50%

C. 38%

D. 22%

5) A sports team held a fundraiser by selling T-shirts and wristbands. They sold T-shirts for $15 each and wristbands for $2 each. They sold a total of 200 items, out of which 120 were T-shirts and the remaining were wristbands. How much money did the team raise from selling wristbands?

A. $120

B. $110

C. $80

D. $160

6) A factory produced 4,500 units of a product in a week. If each unit weighs approximately 2.5 pounds, which of the following is closest to the total weight of the units produced in that week?

A. 5,500 *lbs*

B. 8,000 *lbs*

C. 9,500 *lbs*

D. 11,250 *lbs*

7) During a probability experiment, Sarah flipped three coins. The outcomes of the first 50 trials are shown in the table below:

Probability Experiment	
Faces Showing on	Number of Flipped Coins Outcomes
3 tails	7
1 head, 2 tails	14
2 heads, 1 tail	18
3 heads	11

Based on the information in the table, in how many of the next 150 trials will the outcome be at least one head and one tail?

A. 90

B. 103

C. 96

D. 150

8) The cost of a bicycle is $350. The sales tax rate is 8%. What is the sales tax on this bicycle in dollars and cents?

A. $24.50

B. $28.00

C. $27.75

D. $23.50

9) The meteorologist predicts a $\frac{3}{5}$ chance of precipitation for Saturday. How can we describe the likelihood of rain on this day?

   A. It is extremely unlikely that it will rain on Saturday.

   B. There is a minimal chance of rain on Saturday.

   C. It is quite likely that it will rain on Saturday.

   D. It is absolutely certain that it will rain on Saturday.

10) The circumference of a circle is 62.8 inches. What is the area in square inches? Use 3.14 for π.

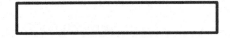

11) A company plans to ship 5,000 bottles of shampoo. The company randomly selects 200 bottles and finds that twelve bottles have a defective cap. Based on this data, how many bottles out of the 5,000 should be predicted to have a defective cap?

   A. 250

   B. 60

   C. 425

   D. 300

12) Sarah is walking in the park. She completes 60% of her walk in $2\frac{1}{4}$ hours. She continues walking at that same rate. How much time, in hours, will Sarah's entire walk take?

   A. $3\frac{1}{2}$

   B. 4

   C. $3\frac{3}{4}$

   D. 5

13) The table below shows the numbers of pages Maria can read in certain amounts of time.

Maria's Reading Rate	
Time (minutes)	Pages Read
45	60
60	80
75	100
90	120

Based on the table, which equation can be used to determine the number of pages (p) Maria can read in t minutes?

A. $p = \left(\frac{4}{3}\right)t$

B. $p = \left(\frac{3}{4}\right)t$

C. $t = \left(\frac{4}{3}\right)p$

D. $t = \left(\frac{3}{4}\right)p$

14) A stone with a triangular face is used in a water fountain. Tiffany made a scale drawing of the triangular face. She labels the vertices of her triangle as P, Q, and R. The sides of triangle PQR are described below.

- Side PQ is 16 inches long.

- Side QR is 20% longer than side PQ.

- Side PR is $\frac{3}{4}$ the length of side QR.

Each inch of Tiffany's scale drawing represents $\frac{1}{4}$ foot of the actual triangular face. What is the perimeter, in feet, of the actual triangular face of the stone?

A. 16

B. 12.4

C. 20.5

D. 24.5

15) A square pyramid is cut into two pieces with a single straight cut. The cut passes through the vertex of the pyramid and is perpendicular to the base. What shape is created by the cross section of the cut?

A. a square

B. a right triangle

C. a rectangle that is not a square

D. an isosceles triangle

16) 200 people were asked about their favorite type of music. The data collected is shown in the graph below.

How much is the central angle in "Jazz"?

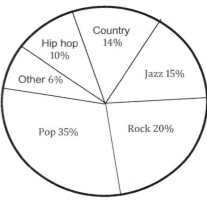

17) Which equation represents the relationship between x and y in the graph?

A. $y = 2x - 3$

B. $y = 3x - 2$

C. $y = -2x - 3$

D. $y = x + 2$

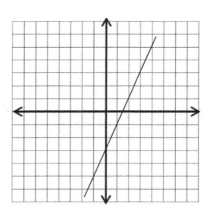

18) A company has four production facilities (A, B, C, and D) across the country. Each facility employs over 200 workers. The company wants to determine the most effective safety training program. Which sample should the company use to arrive at the most reliable conclusion?

    A. 20 employees from facility A

    B. 20 employees from each facility

    C. 50 employees from facility A

    D. 50 employees from each facility

19) The dimensions of a cylindrical tank are 10 feet in diameter and 20 feet high. The tank is to be filled with water up to 90% capacity. Each cubic foot contains 7.48 gallons of water. How many gallons of water, to the nearest tenth, are needed to fill the tank to 90% capacity?

    A. 1,570

    B. 1,413

    C. 10,569.24

    D. 11,265.46

20) Ms. Johnson randomly selects a student from her biology class each day. Each student is equally likely to be selected. There is an equal number of freshmen, sophomores, juniors, and seniors in her class. On Monday, Tuesday, Wednesday, and Thursday of this week, the randomly selected student is a sophomore. Which statement best describes the probability Ms. Johnson selects a sophomore on Friday?

    A. The probability Ms. Johnson selects a sophomore on Friday is the same as it was on each of the other days.

    B. The probability Ms. Johnson selects a sophomore on Friday is less than it was on other days because she has already selected a sophomore 4 days in a row.

    C. The probability Ms. Johnson selects a sophomore on Friday is greater than it was on other days because she has already selected a sophomore 4 days in a row.

    D. The probability Ms. Johnson selects a sophomore on Friday is impossible to determine without knowing how many students are in her class.

21) The figure below consists of a triangle, a rectangle and a semicircle. What is the total area of the shape?

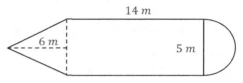

A. $102.56\ m^2$

B. $85\ m^2$

C. $79.81\ m^2$

D. $94.81\ m^2$

22) The related information of the two data series is shown in the box plots below. Which statement is best supported by the information in the box plots?

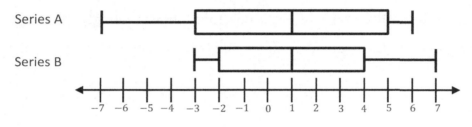

hours students spent studying

A. The range of series B data is larger than the range of series A data.

B. The interquartile range of series A data is equal to the interquartile range of series B data.

C. The median of data in series A is equal to the median of data in series B.

D. The smallest number in the data of series A is equal to the smallest number in the data of series B.

23) The ratio of cats to dogs in Mrs. Smith's animal shelter is 5 to 7. If there are 35 cats in the shelter, how many dogs are there?

24) A roller coaster is descending down a hill at a constant rate. The equation $y = -2x + 10$ can be used to represent this situation, where y is the height of the roller coaster in meters above the ground and x is the distance the roller coaster has traveled in meters.

Which statement best describes the height of the roller coaster, given this equation?

A. From a starting height of 10 meters above the ground, the roller coaster is descending 2 meters per meter traveled.

B. From a starting height of 10 meters above the ground, the roller coaster is ascending 2 meters per meter traveled.

C. From a starting height of 2 meters above the ground, the roller coaster is descending 10 meters per meter traveled.

D. From a starting height of 2 meters above the ground, the roller coaster is ascending 10 meters per meter traveled.

25) Katie has a collection of 900 marbles. Of these marbles, 30% are red, 50% are blue, and the rest are green.

How many marbles in Katie's collection are green?

A. 90

B. 180

C. 270

D. 360

26) Which equation is true when $x = 5$?

A. $5x + 1 = 19$

B. $x - 2 = 7$

C. $3x + 7 = 22$

D. $4x - 4 = 14$

27) Helen biked for 30 minutes at a speed of $10\frac{km}{h}$, then jogged for 1 hour at a speed of $5\frac{km}{h}$, and then swam for 45 minutes at a speed of $2\frac{km}{h}$. What was the total distance Helen covered during this time?

A. $11.5\ km$

B. $8.5\ km$

C. $9.5\ km$

D. $10.5\ km$

28) Maggie is making cupcakes for a school bake sale. She needs to make 120 cupcakes, and she plans to use 2 cups of flour for every 12 cupcakes. Maggie has 10 cups of flour in her pantry. How many more cups of flour does Maggie need to buy in order to have enough flour for all the cupcakes she plans to make?

A. 10, because $\left(\frac{120}{12}\right) \times 2 - 10 = 10$

B. 1, because $\left(\frac{120}{12}\right) \times 2 \div 10 = 1$

C. 7, because $\left(\frac{120}{12}\right) \times 10 + 2 = 30; 4$

D. 30, because $\left(\frac{120}{12}\right) \times 2 + 10 = 30$

29) Ms. Johnson sold 100 T-shirts and 75 hats in 2 days. If she continues selling at the same rate, how many more T-shirts than hats will she sell in 6 days?

A. 65

B. 75

C. 125

D. 100

30) Which two statistical measures are most appropriate for describing the central tendency and variability of the following data set, which represents the daily sales (in dollars) of a retail store for the past 12 days:

$345, $410, $335, $490, $285, $380, $410, $390, $360, $420, $395, $410?

A. Mean and median

B. Mean and standard deviation

C. Mode and median

D. Mode and range

31) What is the result of the following expression: $32.25 \div \left(4\frac{2}{5}\right)$?

A. 8.06

B. 8.64

C. 3.56

D. 7.32

32) A school is planning a field trip for its students. The cost of admission is $20 per student and the cost of renting a bus is $500. If s represents the number of students going on the field trip, which inequality represents all possible values of s, the number of students that can go on the field trip with a budget of $1,500?

A. $20s + 500 \leq 1,500$

B. $20s + 500 \geq 1,500$

C. $20s + 500 < 1,500$

D. $20s + 500 > 1,500$

33) What is the answer to the inequality $3x + 5 \leq 26$?

  A. $x \leq \frac{31}{3}$

  B. $x \leq 6$

  C. $x \geq 7$

  D. $x \leq 7$

34) Which of the following sequences is an arithmetic progression with a common difference of 2 and a first term of 9?

  A. $9,11,13,17,19,...$

  B. $9,-7,3,-1,5,...$

  C. $9,5,7,11,13,...$

  D. $9,6,3,0,-3,...$

35) What is the median of these numbers? $1,11,13,26,3,19,31,9$

  ┌─────────────────────┐
  │   │
  └─────────────────────┘

36) Find the value of x according to the following model.

  A. $x = \frac{1}{5}$

  B. $x = -\frac{1}{5}$

  C. $x = -\frac{1}{7}$

  D. $x = -\frac{2}{5}$

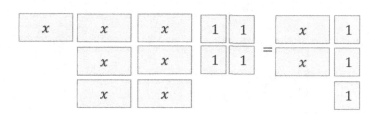

37) If p is directly proportional to the cube of q, and $q = 3$ when $p = 81$, then when $p = 192$, $q =$?

A. 4

B. 5

C. 6

D. 7

38) The price of a laptop was decreased from \$800.00 to \$680.00. By what percentage was the price of the laptop decreased?

A. 12.5%

B. 13%

C. 17.5%

D. 15%

39) If $\frac{3}{7}$ of a tank contains 168 liters of water, what is the capacity of six such tanks of water?

A. 840

B. 2,352

C. 1,260

D. 1,933

40) If $-4 \leq x < 1$, what is the minimum value of the following expression?

$$5x + 3$$

A. -17

B. 8

C. -20

D. 0

End of STAAR Grade 7 Math Practice Test 3

STAAR Mathematics

Practice Test 4

2024

Grade 7

Total number of questions: 40

Total time to complete the test: No time limit

You may NOT use a calculator.

45

STAAR Grade 7 Mathematics Formula Sheet

LINEAR EQUATIONS

Slope – intercept form $y = mx + b$

Direct Variation $y = kx$

Slope of a Line

$$m = \frac{y_2 - y_1}{x_2 - x_1}$$

CIRCUMFERENCE

Circle $C = 2\pi r$ or $C = \pi d$

AREA

Triangle

$$A = \frac{1}{2}bh$$

Parallelogram $A = bh$

Trapezoid

$$A = \frac{1}{2}h(b_1 + b_2)$$

Circle $A = \pi r^2$

SURFACE AREA

	Lateral	Total
Prism	$S = Ph$	$S = Ph + 2B$
Cylinder	$S = 2\pi rh$	$S = 2\pi rh + 2\pi r^2$

VOLUME

Prism or Cylinder $V = Bh$

Pyramid or Cone

$$V = \frac{1}{3}Bh$$

Sphere

$$V = \frac{4}{3}\pi r^3$$

ADDITIONAL INFORMATION

Pythagorean theorem $a^2 + b^2 = c^2$

Simple interest $I = prt$

Compound Interest $A = p(1 + r)^t$

1) The weather forecast for Saturday shows a 48% chance of rain. How would you describe the probability of precipitation on this day?

   A. It is highly unlikely that it will rain on Saturday.

   B. There is an almost even chance of rain on Saturday.

   C. It is likely that it will rain on Saturday.

   D. Rain is guaranteed on Saturday.

2) The circumference of a circular swimming pool is 37.68 meters. What is the area, in square meters, of the pool? Use 3.14 for π.

   A. $6\ m^2$

   B. $31.4\ m^2$

   C. $103\ m^2$

   D. $113.04\ m^2$

3) A stone with a triangular face is used in a water fountain. Emma made a scale drawing of the triangular face. She has labeled the vertices of her triangle as P, Q, and R. The sides of triangle PQR are described below.
   • Side PQ is 20 inches long.
   • Side QR is 30% longer than side PQ.
   • Side PR is $\frac{1}{2}$ the length of side QR.

   Each inch of Emma's scale drawing represents $\frac{1}{3}$ foot of the actual triangular face. What is the perimeter, in feet, of the actual triangular face of the stone?

   A. 14

   B. 19.7

   C. 22.5

   D. 28

4) What is the equation that represents the linear relationship between the x −values and the y −values in the given table?

A. $y = 5x - 3$

B. $y = 5x + 3$

C. $y = 5x$

D. $y = 5x - 5$

x	y
0	−3
3	12
4	17
7	32

5) A teacher gave a test to her class, and the scores are as follows: 75, 82, 60, 90, 92, 68, 80, 77, 85, and 72. Which measure of central tendency would be most useful to describe the performance of the class on this test?

A. Mean

B. Median

C. Mode

D. Range

6) The following income statement represents the revenues and expenses of a small business. However, the cost of goods sold (COGS) is not given.

Revenue	Amount
Sales	$50,000

Expenses	Amount
Rent	$5,000
Utilities	$2,500
Payroll	$25,000
COGS	?

If the net income of the business is $10,000, what is the cost of goods sold?

A. $15,000

B. $17,500

C. $7,500

D. $22,500

7) A company is conducting a survey to determine the preferences of its employees for an upcoming team-building event. Out of 50 employees, 10 were randomly selected for the survey. The results of the survey are as shown.

• 5 employees prefer a sports event.

• 3 employees prefer a board game event.

• 2 employees prefer a cooking event.

Based on the survey results, which of these is the best prediction of the preferences of all 50 employees for the team-building event?

A. There are 25 employees who prefer a sports event.

B. There are 15 employees who prefer a board game event.

C. There are 10 more employees who prefer a sports event than those who prefer a board game event.

D. There are 3 more employees who prefer a board game event than those who prefer a cooking event.

8) Sarah has a collection of 800 books. Of these books, 20% are biographies, 40% are fiction, and the rest are non-fiction. How many books in Sarah's collection are non-fiction?

A. 160

B. 240

C. 320

D. 480

9) What is the area of the shaded region?

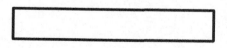

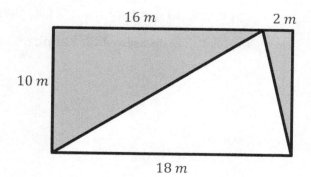

10) What are the values of d that make the inequality $-5d + 5\frac{1}{2} \le 17$ true?

   A. $d \le -2\frac{3}{10}$

   B. $d \ge -2\frac{3}{10}$

   C. $d \ge -\frac{3}{10}$

   D. $d \le \frac{3}{10}$

11) The net of the pyramid is shown in the diagram. What is its surface area?

   A. $120 \; cm^2$

   B. $288 \; cm^2$

   C. $64 \; cm^2$

   D. $224 \; cm^2$

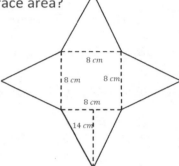

12) Which geometric sequence is represented by the expression 2^{x-1}, where x represents the position of a term in the sequence?

   A. $1, 2, 4, 8, 16, \ldots$

   B. $2, 4, 8, 16, 32, \ldots$

   C. $\frac{1}{2}, 1, 2, 4, 8, \ldots$

   D. $16, 8, 4, 2, 1, \ldots$

13) A restaurant serves pizza with 4 different toppings: pepperoni, mushrooms, olives, and onions. The percentages of customers who order each topping are as follows: 45% pepperoni, 30% mushrooms, 20% olives, and 5% onions. Sarah orders a pizza at random. What is the probability that her pizza does not have olives on it?

   A. 0.05

   B. 0.20

   C. 0.45

   D. 0.8

14) Mr. Patel randomly selects a customer from his store each day. Each customer is equally likely to be selected. There is an equal number of male and female customers in his store. On Monday, Tuesday, Wednesday, and Thursday of this week, the randomly selected customer is a female. Which statement best describes the probability Mr. Patel selects a female customer on Friday?

A. The probability Mr. Patel selects a female customer on Friday is the same as it was on each of the other days.

B. The probability Mr. Patel selects a female customer on Friday is less than it was on other days because he has already selected a female customer 4 days in a row.

C. The probability Mr. Patel selects a female customer on Friday is greater than it was on other days because he has already selected a female customer 4 days in a row.

D. The probability Mr. Patel selects a female customer on Friday is impossible to determine without knowing how many customers are in his store.

15) The table below shows the numbers of pages Alex can read in certain amounts of time.

Alex's Reading Rate	
Time (minutes)	Pages Read
20	30
30	45
40	60
50	75

Based on the table, which equation can be used to determine the number of pages (p) Alex can read in t minutes?

A. $p = \left(\frac{2}{3}\right)t$

B. $t = \left(\frac{3}{2}\right)p$

C. $p = \left(\frac{3}{2}\right)t$

D. $t = \left(\frac{4}{3}\right)p$

16) A company plans to ship 10,000 boxes of cookies. The company randomly selects 500 boxes and finds that twenty boxes have broken seals. Based on this data, how many boxes out of the 10,000 should be predicted to have a broken seal?

A. 20

B. 400

C. 200

D. 350

17) A college student has a monthly budget as follows:

Budget	
Items	Amount
Rent	$500
Textbooks	$200
Food	$250
Transportation	$100
Entertainment	$150
Miscellaneous	$50

Which equation can be used to find the minimum amount of money the student must earn annually in order to meet this budget?

A. $b = \$1,250 \times 12$

B. $b = \$1,250 \times 52$

C. $b = \$2,300 \div 52$

D. $b = \$2,300 \div 365$

18) A recipe calls for $\frac{3}{5}$ cups of flour for every $\frac{2}{3}$ teaspoon of salt. What is the unit rate of cups per teaspoon?

A. $\frac{2}{5}$

B. $\frac{3}{10}$

C. $\frac{2}{10}$

D. $\frac{9}{10}$

19) A spherical balloon has a radius of 2 feet. The balloon is to be filled with helium up to 75% capacity. Each cubic foot contains 0.0118 pounds of helium. How many pounds of helium, to the nearest hundredth, are needed to fill the balloon to 75% capacity?

A. 0.354

B. 0.312

C. 0.200

D. 0.296

20) A school has a budget of $10,000 to purchase textbooks and workbooks for their students. The textbooks cost $50 each and the workbooks cost $10 each. Let t represent the number of textbooks and w represent the number of workbooks. Which inequality represents all possible values of t and w, the number of textbooks and workbooks they can buy with the given budget?

A. $50t + 10w \leq 10,000$

B. $50t + 10w \geq 10,000$

C. $10t + 50w < 10,000$

D. $10t + 50w > 10,000$

21) A retail store chain wants to determine the most effective advertising campaign. There are ten stores in the chain, and each store has over 50 employees. Which sample should the company use to arrive at the most reliable conclusion?

A. 5 employees from each store

B. 10 employees from one store

C. 25 employees from each store

D. 50 employees from each store

22) A bag contains 5 red, 5 blue, 5 green, and 5 yellow marbles. A marble is randomly pulled from the bag and replaced eight times. The table shows the outcome of the experiment.

Which marble color's observed frequency is closest to its expected frequency?

A. Red

B. Green

C. Blue

D. Yellow

Trial Outcome	
1	Green
2	Red
3	Yellow
4	Blue
5	Green
6	Blue
7	Yellow
8	Green

23) According to the following figure, which point is inside the circle?

A. $(3, -2)$

B. $(0, -2)$

C. $(-1, 1)$

D. $(-3, 3)$

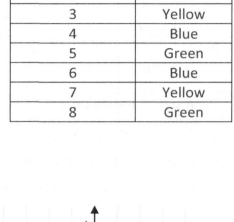

24) A map of a city has a scale of 1 inch represents 2 miles. If the actual distance between two locations on the map is 8 miles, what is the distance between these locations on the map, in inches?

A. 2 inches.

B. 4 inches.

C. 8 inches.

D. 16 inches.

25) Hannah has a basket with 8 oranges, 3 apples, and 5 pears. If Hannah randomly selects one fruit from the basket, what is the probability that she will choose an apple?

A. $\frac{5}{16}$

B. $\frac{1}{2}$

C. $\frac{3}{15}$

D. $\frac{3}{16}$

26) In a bakery, the ratio of chocolate to vanilla cupcakes is 3 to 5. If there are 80 cupcakes in the bakery, how many chocolate cupcakes are there?

27) What is the slope of a line that is parallel to the line with equation of $y - 5x = 4$?

A. 5

B. 1

C. -5

D. 4

28) The base of a triangle is 4 meters and its height is 3 meters. Which equation can be used to determine A, the area of this triangle in square centimeters?

A. $\frac{1}{2} \times 4 \times 3 \; cm^2$

B. $\frac{1}{2} \times 400 \times 300 \times 10 \; cm^2$

C. $\frac{\frac{1}{2} \times 400 \times 300}{10} \; cm^2$

D. $100 \, (600) \; cm^2$

29) The length of a rectangular pool in a scale drawing is 8 inches and the actual length is 40 feet. If the width of the pool in the scale drawing is 5 inches, what is the actual width of the pool?

30) If you are driving at an average speed of 70 miles per hour on a journey that is 490 miles long, how many hours will it take you to arrive at your destination?

A. 7 hours

B. 6 hours and 30 minutes

C. 7 hours and 30 minutes

D. 8 hours and 30 minutes

31) Bob has a savings account with a balance of $10,000. The bank offers an interest rate of 2.5% per year on the savings. If Bob doesn't make any withdrawals or deposits for one year, how much interest will he earn on his savings account at the end of the year?

A. $200.50

B. $250.00

C. $225.00

D. $275.00

32) Lila has a monthly income of $5,000. She spends a portion of her income on different expenses as listed in the table below. What percentage of her income is used for her transportation and entertainment expenses?

A. 13%

B. 14%

C. 15%

D. 16%

Monthly Expenses	
Category	Amount of Money
Rent	$1,500
Utilities	$200
Groceries	$450
Transportation	$350
Entertainment	$300

33) The image of the figure below consists of two trapezoids and a semicircle. What is the total area of the shape?

A. 139 m^2

B. 69.13 m^2

C. 98.13 m^2

D. 153.13 m^2

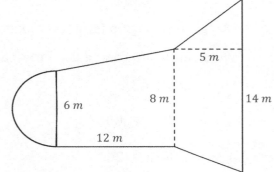

34) A rectangular swimming pool measures 25 feet by 50 feet. If there are approximately 0.3048 meters in 1 foot, which measurement is closest to the area of the pool in square meters?

A. 99.2 m^2

B. 106.15 m^2

C. 116.13 m^2

D. 133.32 m^2

35) Triangle ABC is an isosceles triangle. The measure of angle A is 70°. The measure of angle B is $(3x + 10)$°. Which equation can be used to find the value of x?(The measure of angle C is equal to the measure of angle B)

A. $70 = 3x + 10$

B. $70 + (3x + 10) = 180$

C. $6x + 90 = 180$

D. $70 + 3 \times (3x + 10) = 180$

36) Which of the following graphs represents the compound inequality $-3 < 4x - 3 \leq 1$?

A.

B.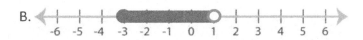

C.

D.

37) Maggie is saving up for a trip to Japan. She starts with $1,000 and saves $200 each month. Which equation can be used to find y, the total amount of money Maggie will have saved after x months?

A. $y = 200x + 1,000$

B. $y = 1,000x + 200$

C. $y = 200x - 1000$

D. $y = 1,000 - 200x$

38) What is the probability of selecting a heart from a standard 52-card deck?

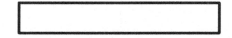

39) A company manufactures 500 toys using 3 bags of stuffing material. If the company needs to manufacture 750 toys, how many bags of stuffing material are required?

A. 7

B. 5

C. 6

D. 8

40) A company has a production target of 200 units per day. On Monday, they produced 150 units, and on Tuesday they produced 175 units, while on Wednesday they produced 225 units. If they maintain the same production rate, how many units will they produce in 5 days?

A. 902.33 units

B. 910.32 units

C. 916.67 units

D. 900 units

End of STAAR Grade 7 Math Practice Test 4

STAAR Mathematics

Practice Test 5

2024

Grade 7

Total number of questions: 40

Total time to complete the test: No time limit

You may NOT use a calculator.

61

STAAR Grade 7 Mathematics Formula Sheet

LINEAR EQUATIONS

Slope – intercept form	$y = mx + b$
Direct Variation	$y = kx$
Slope of a Line	$m = \dfrac{y_2 - y_1}{x_2 - x_1}$

CIRCUMFERENCE

Circle	$C = 2\pi r$ or $C = \pi d$

AREA

Triangle	$A = \dfrac{1}{2} bh$
Parallelogram	$A = bh$
Trapezoid	$A = \dfrac{1}{2} h(b_1 + b_2)$
Circle	$A = \pi r^2$

SURFACE AREA

	Lateral	Total
Prism	$S = Ph$	$S = Ph + 2B$
Cylinder	$S = 2\pi rh$	$S = 2\pi rh + 2\pi r^2$

VOLUME

Prism or Cylinder	$V = Bh$
Pyramid or Cone	$V = \dfrac{1}{3} Bh$
Sphere	$V = \dfrac{4}{3} \pi r^3$

ADDITIONAL INFORMATION

Pythagorean theorem	$a^2 + b^2 = c^2$
Simple interest	$I = prt$
Compound Interest	$A = p(1 + r)^t$

1) The table below shows the number of pages David can read in various amounts of time.

David's Reading Rate	
Time (minutes)	Pages Read
30	40
45	60
60	80
75	100

Based on the table, which equation can be used to determine the number of pages (p) David can read in t minutes?

A. $p = \left(\frac{4}{3}\right)t$

B. $p = \left(\frac{3}{4}\right)t$

C. $t = \left(\frac{4}{3}\right)p$

D. $t = \left(\frac{3}{4}\right)p$

2) What is the equation of the line shown in the graph below?

A. $y = -\frac{5}{11}x + 10$

B. $y = \frac{11}{10}x + 9$

C. $y = -\frac{10}{11}x + 10$

D. $y = -\frac{7}{10}x + 11$

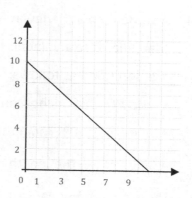

3) The net worth statement below shows Jack's financial situation. Assets are represented with positive values, and liabilities are shown with negative values. However, the current value of his jewelry is not given.

Item	Value
Jewelry (current value)	
Savings account	$4,000
Credit-card debt	−$1,500
Stocks	$8,500
Student loans	−$25,000
IRA	$20,000
Car loan	−$7,500

If Jack's net worth is $15,000, what is the current value of Jack's jewelry?

A. $12,500

B. $16,500

C. $15,500

D. $13,000

4) A stone with a triangular face is used in a water fountain. Jason made a scale drawing of the triangular face and labels the vertices of his triangle as P, Q, and R. The sides of triangle PQR are described as follows:

• Side PQ is 18 inches long.

• Side QR is 10% longer than side PQ.

• Side PR is $\frac{2}{3}$ the length of side QR. Each inch of Jason's scale drawing represents $\frac{1}{8}$ foot of the actual triangular face. What is the perimeter, in feet, of the actual triangular face of the stone?

A. 7.635

B. 4.725

C. 2.75

D. 6.375

5) Joe cycled for 2 hours at a speed of $10\frac{km}{h}$, then walked for 1 hour at a speed of $4\frac{km}{h}$, and then cycled again for 1.5 hours at a speed of $12\frac{km}{h}$. What was the total distance Joe covered during this time?

6) Which expression is represented by the model below?

A. $4.(5)$

B. $4.(-5)$

C. $5.(-5)$

D. $(4).(1)$

7) Nina collected data on the height and weight of four different individuals. The information she gathered is shown in the table below:

Height and Weight Comparison		
Name	Height (inches)	Weight (pounds)
A	68	170
B	72	200
C	65	150
D	70	180

Based on the data in the table, which individual has the highest weight-to-height ratio?

A. Individual A

B. Individual B

C. Individual C

D. Individual D

8) The total bill for a restaurant meal is $75, and the tip rate is 15%. How much should be left as a tip in dollars and cents?

A. $10.50

B. $10.25

C. $11.50

D. $11.25

9) A survey was conducted to determine the preferred social media platforms among 400 teenagers. The results were displayed in a pie chart, where "Instagram" had the largest slice. What is the central angle in degrees that represents the "Instagram" slice in the pie chart?

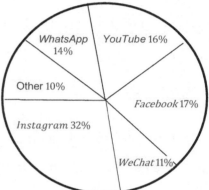

A. 102.3°

B. 99.8°

C. 115.2°

D. 121.1°

10) The weather forecast indicates a $\frac{3}{4}$ probability of rain on Saturday. What can we infer about the chances of rainfall on that day?

A. There is no chance of rain on Saturday.

B. Rainfall on Saturday is highly probable.

C. The likelihood of rain on Saturday is moderate.

D. It is very unlikely to rain on Saturday.

11) A university aims to determine the effectiveness of its career counseling services. There are five schools (engineering, business, education, arts and sciences, and health sciences) at the university, and each school has over 1,000 students. Which sample should the university use to reach the most reliable conclusion?

   A. 10 students from each school

   B. 20 students from one school

   C. 50 students from each school

   D. 100 students from each school

12) A pizza parlor maintains a ratio of 3 pepperoni pizzas to 5 cheese pizzas. If there are 24 pepperoni pizzas, how many cheese pizzas are there?

   ┌─────────────────────────┐
   │   │
   └─────────────────────────┘

13) Which of the following equations is true when $x = 3$?

   A. $2x + 3 = 9$

   B. $x + 4 = 8$

   C. $5x - 2 = 11$

   D. $3x - 7 = 0$

14) A rectangular fish tank measures 3 feet by 2 feet by 2 feet. The tank is to be filled with water up to 60% of its capacity. Each cubic foot holds 7.48 gallons of water. How many gallons of water, to the nearest tenth, are required to fill the tank to 60% capacity?

   A. 48.33 gallons

   B. 7.2 gallons

   C. 53.86 gallons

   D. 59.24 gallons

15) If $\frac{5}{8}$ of a swimming pool can accommodate 2,400 gallons of water, what is the total capacity of three such swimming pools?

A. 10,100 *gallons*

B. 11,520 *gallons*

C. 12,340 *gallons*

D. 7,330 *gallons*

16) The following question relates to a set of data on the number of hours per week that a group of students spend studying.

Given the data in the table below, what is the range of the data and what is the median?

A. Range: 5; Median: 6

B. Range: 6; Median: 7

C. Range: 5; Median: 5.5

D. Range: 5; Median: 6.5

Student	Hours
A	4
B	5
C	5
D	6
E	7
F	8
G	9

17) A toy manufacturer has a budget of $3,000 to produce two types of toys: type A and type B. Type A toys cost $50 each to produce, and type B toys cost $30 each to produce. Let a represent the number of type A toys and b represent the number of type B toys. Which inequality represents all possible values of a and b, i.e., the number of each type of toy they can produce with the given budget?

A. $50a + 30b \leq 3,000$

B. $50a + 30b \geq 3,000$

C. $50b + 30a < 3,000$

D. $50b + 30a > 3,000$

18) The circumference of a circular table is 18.84 feet. What is the area in square feet? Use 3.14 for π.

    A. 26.33 *square feet*

    B. 28.26 *square feet*

    C. 33.78 *square feet*

    D. 29.56 *square feet*

19) A company intends to ship 3,000 packages of toys. The company randomly selects 150 packages and finds that eight packages have a missing item. Based on this data, how many packages out of the 3,000 are predicted to have a missing item?

    A. 150

    B. 180

    C. 140

    D. 160

20) Sophie has 4 red balls, 3 green balls, and 2 blue balls in a box. If she randomly selects one ball from the box, what is the probability that she will select a green ball?

21) Which of the following sequences is an arithmetic progression with a common difference of 3 and a first term of 7?

    A. 7, 10, 13, 16, 19, ...

    B. 7, 13, 19, 25, 31, ...

    C. 7, −4, −15, −26, −37, ...

    D. 7, 4, 1, −2, −5, ...

22) Suppose the manager of a fast-food restaurant surveyed 100 customers who visited the restaurant on a Monday. The manager asked each customer about their favorite item on the menu. The results of the survey are as follows:

Based on the survey results, which of the following statements about a person who will visit the restaurant on the next Monday is true?

Survey Results	
Favorite Item	Number of People
Burger	45
Fries	25
Pizza	10
Chicken	20

A. The person is twice as likely to prefer a burger as fries.

B. The person is four times as likely to prefer pizza as chicken.

C. The person is less likely to prefer chicken or pizza than fries.

D. The person is more likely to prefer burger or chicken than fries or pizza.

23) A car is accelerating from rest at a constant rate. The equation $v = 5t + 2$ can be used to represent this situation, where v is the velocity of the car in meters per second and t is the time elapsed in seconds.

Which statement best describes the velocity of the car, given this equation?

A. From a starting velocity of 2 meters per second, the car is accelerating at a rate of 5 meters per second per second.

B. From a starting velocity of 2 meters per second, the car is decelerating at a rate of 5 meters per second per second.

C. From a starting velocity of 5 meters per second, the car is accelerating at a rate of 2 meters per second per second.

D. From a starting velocity of 5 meters per second, the car is decelerating at a rate of 2 meters per second per second.

24) What is the value of x in the following figure? (Figure not drawn to scale)

A. 8

B. 10

C. 11

D. 12

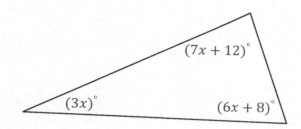

25) A fashion designer created a sketch of a dress using a scale in which 1 centimeter represents 2 inches. The actual length of the dress is 60 inches. What is the length of the dress in the sketch, in centimeters?

A. 30 cm

B. 60 cm

C. 120 cm

D. 240 cm

26) An airplane covers a distance of 1,600 kilometers in 4 hours, while a helicopter covers a distance of 800 kilometers in 2 hours. What is the ratio of the average speed of the airplane to the average speed of the helicopter?

A. $2:1$

B. $1:2$

C. $1:1$

D. $4:3$

27) If a company recorded the number of products sold in a week as follows: $15, 20, 25, 30, 35$, what statistical measure would best describe the variability of the sales data?

A. Mean

B. Median

C. Mode

D. Range

28) Which of the following is equivalent to $-18 < 5x + 7 < 12$

    A. $-3 < x < 4$

    B. $-1 < x < 5$

    C. $-5 < x < 1$

    D. $5 < x < 10$

29) If Mr. Rodriguez sold 51 cell phones and 102 laptops in 3 days and maintains the same selling pace, how many more laptops than cell phones will he sell in 9 days?

    A. 153

    B. 207

    C. 150

    D. 246

30) If y is directly proportional to the fourth power of x, and $x = 2$ when $y = 2$, then when $y = 512, x$ equals?

    A. 8

    B. 5

    C. 6

    D. 7

31) A carpenter needs 5,500 cm^2 of wood to make 50 identical chairs. What is the amount of wood needed in square centimeters to make one chair?

32) David needs to achieve an average of 80% in his physics class to qualify for an advanced placement program. In his first four tests, he received scores of 78%, 81%, 85%, and 88%. What is the minimum score David must obtain on his fifth and final test to qualify for the advanced placement program?

A. 67%

B. 69%

C. 68%

D. 70%

33) John has completed 18 out of 25 problems in his physics homework. What percentage of the problems has John not completed?

A. 33%

B. 17%

C. 46%

D. 28%

34) If $3x - 2y = 6$ and $5x + 4y = 10$, which of the following ordered pairs (x, y) satisfies both equations?

A. (1,2)

B. (2,1)

C. (0,2)

D. (2,0)

35) What is the volume of the following cylinder?

A. $753.6 \ m^3$

B. $188.4 \ m^3$

C. $2,826 \ m^3$

D. $698.5 \ m^3$

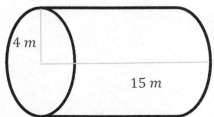

4 m

15 m

36) Using the information provided, what is the total net worth of Sarah's assets and liabilities?

Asset and Liability Statement	
Assets	**Value**
Savings account	$15,000
Investments	$25,000
House (current value)	$300,000
Car	$20,000
Total Assets	**$360,000**
Liabilities	**Value**
Mortgage	$200,000
Car loan	$10,000
Credit card debt	$5,500
Total Liabilities	**$215,500**

A. $360,000

B. $144,500

C. $215,500

D. $144,000

37) Triangle ABC and triangle EFG are similar. Which proportion can be used to find the length of FG?

A. $\dfrac{22}{21} = \dfrac{14}{FG}$

B. $\dfrac{31}{22} = \dfrac{19.7}{FG}$

C. $\dfrac{21}{22} = \dfrac{FG}{19.7}$

D. $\dfrac{21}{22} = \dfrac{14}{FG}$

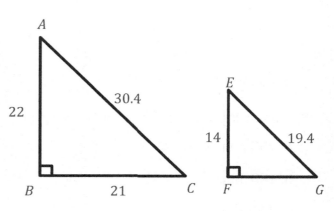

38) What is the result of the following expression: $22.68 \div \left(3\frac{1}{5}\right)$?

   A. 7.56

   B. 72.57

   C. 7.09

   D. 9.24

39) The measure of the angles of a triangle are in the ratio $3:4:8$. What is the measure of the largest angle?

   A. 90°

   B. 12°

   C. 36°

   D. 96°

40) What is the probability of rolling a total of 9 or 12 when two fair dice are rolled at the same time?

   A. $\frac{1}{6}$

   B. $\frac{5}{36}$

   C. $\frac{1}{12}$

   D. $\frac{7}{36}$

End of STAAR Grade 7 Math Practice Test 5

STAAR Mathematics

Practice Test 6

2024

Grade 7

Total number of questions: 40

Total time to complete the test: No time limit

You may NOT use a calculator.

77

STAAR Grade 7 Mathematics Formula Sheet

LINEAR EQUATIONS

Slope – intercept form	$y = mx + b$
Direct Variation	$y = kx$
Slope of a Line	$m = \dfrac{y_2 - y_1}{x_2 - x_1}$

CIRCUMFERENCE

Circle	$C = 2\pi r$ or $C = \pi d$

AREA

Triangle	$A = \dfrac{1}{2}bh$
Parallelogram	$A = bh$
Trapezoid	$A = \dfrac{1}{2}h(b_1 + b_2)$
Circle	$A = \pi r^2$

SURFACE AREA

	Lateral	Total
Prism	$S = Ph$	$S = Ph + 2B$
Cylinder	$S = 2\pi rh$	$S = 2\pi rh + 2\pi r^2$

VOLUME

Prism or Cylinder	$V = Bh$
Pyramid or Cone	$V = \dfrac{1}{3}Bh$
Sphere	$V = \dfrac{4}{3}\pi r^3$

ADDITIONAL INFORMATION

Pythagorean theorem	$a^2 + b^2 = c^2$
Simple interest	$I = prt$
Compound Interest	$A = p(1 + r)^t$

1) The area of a circle is less than 49π. Which of the following can be the circumference of the circle?

   A. 8π

   C. 14π

   B. 16π

   D. 32π

2) A $45 shirt now selling for $27 is discounted by about what percent?

   A. 20%

   B. 37.7%

   C. 40%

   D. 60%

3) From last year, the price of gasoline has increased from $1.40 per gallon to $1.75 per gallon. The new price is what percent of the original price?

   A. 72%

   B. 125%

   C. 140%

   D. 160%

4) If 50% of a class are girls, and only 20% of girls play tennis, what percent of the class play tennis?

   A. 10%

   B. 15%

   C. 20%

   D. 30%

5) What is the median of these numbers? $3, 8, 13, 7, 15, 18, 5$

A. 7

B. 8

C. 13

D. 15

6) Last week 24,000 fans attended a football match. This week three times as many bought tickets, but one sixth of them cancelled their tickets. How many are attending this week?

A. 48,000

B. 54,000

C. 60,000

D. 72,000

7) John traveled 140 km in 7 hours and Alice traveled 180 km in 4 hours. What is the ratio of the average speed of John to average speed of Alice?

A. 3:2

B. 2:3

C. 4:9

D. 5:6

8) What is the value of the expression $3(x - 2y) + (2 - x)^2$ when $x = 5$ and $y = -3$?

A. -22

B. 24

C. 42

D. 88

9) What is the value of x in the following equation? $\frac{2}{3}x + \frac{1}{6} = \frac{1}{3}$

   A. 6

   B. $\frac{1}{2}$

   C. $\frac{1}{3}$

   D. $\frac{1}{4}$

10) A boat sails 12 miles south and then 16 miles east. How far is the boat from its start point?

   A. 18 miles

   B. 20 miles

   C. 24 miles

   D. 28 miles

11) Sophia purchased a sofa for $600. The sofa is regularly priced at $640. What was the percent discount Sophia received on the sofa?

   A. 3.5%

   B. 6.25%

   C. 20%

   D. 25%

12) The score of Emma is half of Ava and the score of Mia is twice that of Ava. If the score of Mia is 40, what is the score of Emma?

   A. 5

   B. 10

   C. 20

   D. 40

13) A bag contains 20 balls: four green, five black, eight blue, a brown, a red, and one white. If 19 balls are removed from the bag at random, what is the probability that a brown ball has been removed?

   A. $\frac{1}{9}$

   B. $\frac{1}{20}$

   C. $\frac{4}{5}$

   D. $\frac{19}{20}$

14) A rope weighs 600 grams per meter of length. What is the weight in kilograms of 15.2 meters of this rope? ($1\ kilograms = 1,000\ grams$)

   A. 0.0912

   B. 0.912

   C. 9.12

   D. 91.20

15) A chemical solution contains 8% alcohol. If there is 38.4 ml of alcohol, what is the volume of the solution?

    A. 240 ml

    B. 480 ml

    C. 600 ml

    D. 1,200 ml

16) If 5 inches on a map represents an actual distance of 100 feet, what actual distance does 18 inches on the map represent?

    A. 18 ft

    B. 100 ft

    C. 250 ft

    D. 360 ft

17) If a gas tank can hold 25 gallons, how many gallons does it contain when it is $\frac{2}{5}$ full?

    A. 50

    B. 125

    C. 62.5

    D. 10

18) Simplify $6x^2y^3(2x^2y)^3 =$?

    A. $12x^4y^6$

    B. $12x^8y^6$

    C. $48x^4y^6$

    D. $48x^8y^6$

19) Right triangle ABC has two legs of lengths $9\ cm$ (AB) and $12\ cm$ (AC). What is the length of the third side (BC)?

   A. $6\ cm$

   B. $8\ cm$

   C. $14\ cm$

   D. $15\ cm$

20) The marked price of a computer is D dollar. Its price decreased by 15% in January and then increased by 10% in February. What is the final price of the computer in D dollar?

   A. $0.805D$

   B. $0.88D$

   C. $0.935D$

   D. $1.20D$

21) What is the value of this expression? $[3 \times (-14) - 48] - (-14) + [3 \times 8] \div 2$

   Write your answer in the box below.

22) The price of a laptop is decreased by 10% to $450. What is its original price?

   A. 320

   B. 380

   C. 400

   D. 500

23) The average of 13, 15, 20 and x is 25. What is the value of x?

   Write your answer in the box below.

24) The ratio of boys and girls in a class is 4 : 7. If there are 55 students in the class, how many more boys should be enrolled to make the ratio 1 : 1?

   A. 8

   B. 10

   C. 12

   D. 15

25) In a party, 14 soft drinks are required for every 16 guests. If there are 160 guests, how many soft drinks is required?

   A. 18

   B. 104

   C. 140

   D. 1440

26) The circle graph below shows all Mr. Green's expenses for last month. If he spent $660 on his car, how much did he spend for his rent?

   A. $700

   B. $740

   C. $780

   D. $810

Mr. Green's monthly expenses

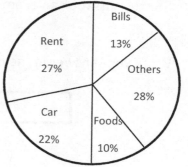

27) 35 is What percent of 20?

    A. 20%

    B. 25%

    C. 175%

    D. 180%

28) Two third of 18 is equal to $\frac{2}{5}$ of what number?

    A. 12

    B. 20

    C. 30

    D. 60

29) The width of a box is one third of its length. The height of the box is one third of its width. If the length of the box is 36 cm, what is the volume of the box?

    A. 81 cm^3

    B. 162 cm^3

    C. 243 cm^3

    D. 1,728 cm^3

30) 55 students took an exam and 11 of them failed. What percent of the students passed the exam?

Write your answer in the box below. (don't write the percent sign.)

31) Two times the price of a laptop is equal to three times the price of a computer. If the price of the laptop is $100 more than the computer, what is the price of the computer?

A. $200

B. $300

C. $600

D. $1200

32) The perimeter of a rectangular yard is 72 meters. What is its length if its width is twice its length? (Don't write the measurement)

Write your answer in the box below.

33) Bob is 12 miles ahead of Mike, who is running at 6.5 miles per hour, and Mike is running at the speed of 8 miles per hour. How long does it take for Bob to catch Mike?

A. 3 hours

B. 4 hours

C. 6 hours

D. 8 hours

34) There are 44 marbles in a jar and 11 of them are red. What percent of the marbles are NOT red?

A. 20%

B. 40%

C. 60%

D. 75%

35) The average weight of 20 girls in a class is $60\ kg$ and the average weight of 30 boys in the same class is $62\ kg$. What is the average weight of all the 50 students in the class?

Write your answer in the box below. (Round your answer to the hundredth place.)

36) A bank is offering 4.5% simple interest on a savings account. If you deposit $9,000, how much interest will you earn in five years?

   A. $405

   B. $720

   C. $2,025

   D. $3,600

37) Six less than twice a positive integer is 68. What is the integer?

   A. 19

   B. 32

   C. 36

   D. 37

38) Which of the following points lies on the line with the equation $3x + 5y = 7$?

   A. $(2, 1)$

   B. $(-1, 2)$

   C. $(-2, 2)$

   D. $(2, 2)$

39) How many possible outfit combinations come from five shirts, seven slacks, and five ties?

Write your answer in the box below.

```
┌─────────────────────────┐
│                         │
└─────────────────────────┘
```

40) An angle is equal to one fourth of its supplement. What is the measure of that angle?

A. 18

B. 24

C. 36

D. 45

End of STAAR Grade 7 Math Practice Test 6

STAAR Mathematics

Practice Test 7

2024

Grade 7

Total number of questions: 40

Total time to complete the test: No time limit

You may NOT use a calculator.

91

STAAR Grade 7 Mathematics Formula Sheet

LINEAR EQUATIONS

Slope – intercept form	$y = mx + b$
Direct Variation	$y = kx$
Slope of a Line	$m = \dfrac{y_2 - y_1}{x_2 - x_1}$

CIRCUMFERENCE

Circle	$C = 2\pi r$ or $C = \pi d$

AREA

Triangle	$A = \dfrac{1}{2}bh$
Parallelogram	$A = bh$
Trapezoid	$A = \dfrac{1}{2}h(b_1 + b_2)$
Circle	$A = \pi r^2$

SURFACE AREA

	Lateral	Total
Prism	$S = Ph$	$S = Ph + 2B$
Cylinder	$S = 2\pi rh$	$S = 2\pi rh + 2\pi r^2$

VOLUME

Prism or Cylinder	$V = Bh$
Pyramid or Cone	$V = \dfrac{1}{3}Bh$
Sphere	$V = \dfrac{4}{3}\pi r^3$

ADDITIONAL INFORMATION

Pythagorean theorem	$a^2 + b^2 = c^2$
Simple interest	$I = prt$
Compound Interest	$A = p(1 + r)^t$

1) Two dice are thrown simultaneously, what is the probability of getting a sum of 5 or 8?

   A. $\frac{1}{3}$

   B. $\frac{1}{4}$

   C. $\frac{1}{16}$

   D. $\frac{11}{36}$

2) A swimming pool holds 2,500 cubic feet of water. The swimming pool is 25 feet long and 10 feet wide. How deep is the swimming pool?

   Write your answer in the box below. (<u>Don't write the measurement</u>)

   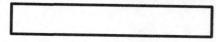

3) Which of the following answers represents the compound inequality $-4 \leq 4x - 8 < 16$?

   A. $-2 \leq x \leq 8$

   B. $-2 < x \leq 8$

   C. $1 < x \leq 6$

   D. $1 \leq x < 6$

4) Which percentage is closest in value to 0.0099?

   A. 2%

   B. 0.1%

   C. 1%

   D. 100%

5) A card is drawn at random from a standard 52–card deck, what is the probability that the card is of Clubs? (The deck includes 13 of each suit clubs, diamonds, hearts, and spades)

   A. $\frac{1}{3}$

   B. $\frac{1}{4}$

   C. $\frac{1}{6}$

   D. $\frac{1}{52}$

6) How long does a 380 − miles trip take moving at 40 miles per hour (mph)?

   A. 9 hours

   B. 9 hours and 24 minutes

   C. 9 hours and 30 minutes

   D. 12 hours and 30 minutes

7) 16 yards 9 feet and 10 inches equal to how many inches?

   A. 169

   B. 561

   C. 694

   D. 961

8) A shirt costing $500 is discounted 25%. After a month, the shirt is discounted another 15%. Which of the following expressions can be used to find the selling price of the shirt?

   A. $(500)(0.60)$

   B. $(500) - 500(0.40)$

   C. $(500)(0.25) - (200)(0.15)$

   D. $(500)(0.75)(0.85)$

9) In the diagram below, circle A represents the set of all odd numbers, circle B represents the set of all negative numbers, and circle C represents the set of all multiples of 5. Which number could be replaced with y?

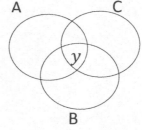

A. 5

B. 0

C. −5

D. −10

10) A juice mixture contains $\frac{5}{14}$ jar of cherry juice and $\frac{5}{70}$ jar of apple juice. How many jars of cherry juice per jar of apple juice does the mixture contain?

Write your answer in the box below.

11) The price of a car was $28,000 in 2012. In 2013, the price of that car dropped to $18,200. What was the rate of depreciation of the price of car per year?

A. 20%

B. 30%

C. 35%

D. 40%

12) What is the equivalent temperature of $140°F$ in Celsius? $C = \frac{5}{9}(F - 32)$

A. 32

B. 40

C. 48

D. 60

13) The square of a number is $\frac{36}{64}$. What is the cube of that number?

   A. $\frac{6}{8}$

   B. $\frac{25}{254}$

   C. $\frac{216}{512}$

   D. $\frac{125}{64}$

14) What is the surface area of the cylinder below?

   A. $28\pi \ in^2$

   B. $37\pi \ in^2$

   C. $40\pi \ in^2$

   D. $288\pi \ in^2$

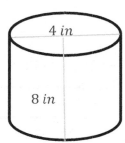

4 in

8 in

15) If $3x - 5 = 8.5$, what is the value of $6x + 3$?

   A. 13

   B. 15.5

   C. 20.5

   D. 30

16) The average of 6 numbers is 12. The average of 4 of those numbers is 10. What is the average of the other two numbers?

   A. 10

   B. 12

   C. 14

   D. 16

17) Anita's trick–or–treat bag contains 15 pieces of chocolate, 10 suckers, 10 pieces of gum, and 25 pieces of licorice. If she randomly pulls a piece of candy from her bag, what is the probability of her pulling out a sucker?

A. $\frac{1}{3}$

B. $\frac{1}{4}$

C. $\frac{1}{6}$

D. $\frac{1}{12}$

18) Which of the following shows the numbers from least to greatest? $\frac{11}{15}, 75\%, 0.74, \frac{19}{25}$

A. $75\%, 0.74, \frac{11}{15}, \frac{19}{25}$

B. $75\%, 0.74, \frac{19}{25}, \frac{11}{15}$

C. $0.74, 75\%, \frac{11}{15}, \frac{19}{25}$

D. $\frac{11}{15}, 0.74, 75\%, \frac{19}{25}$

19) Mr. Carlos' family are choosing a menu for their reception. They have 5 choices of appetizers, 7 choices of entrees, and 3 choices of cake. How many different menu combinations are possible for them to choose?

A. 35

B. 75

C. 105

D. 150

20) Four one – foot rulers can be split among how many users so that each gets $\frac{1}{3}$ of a ruler?

    A. 4

    B. 6

    C. 12

    D. 24

21) What is the area of a square whose diagonal is 4?

    A. 8

    B. 32

    C. 36

    D. 64

22) The ratio of boys to girls in a school is $2:3$. If there are 600 students in total in the school, how many boys are in the school?

    A. 60

    B. 120

    C. 240

    D. 480

23) The diagonal of a rectangle is 13 inches long and the height of the rectangle is 5 inches. What is the area of the rectangle in inches?

    A. 20

    B. 60

    C. 120

    D. 200

24) Mr. Jones saves $5,000 out of his monthly family income of $85,000. What is the fractional part of his income does he save?

A. $\frac{1}{17}$

B. $\frac{1}{13}$

C. $\frac{3}{23}$

D. $\frac{2}{27}$

25) When a number is subtracted from 32 and the difference is divided by that number, the result is 3. What is the value of the number?

A. 2

B. 4

C. 8

D. 12

26) What is the volume of a box with the following dimensions?

$$\text{Height} = 8 \; cm, \text{Width} = 4 \; cm, \text{Length} = 5 \; cm$$

A. $32 \; cm^3$

B. $120 \; cm^3$

C. $140 \; cm^3$

D. $160 \; cm^3$

27) In two successive years, the population of a town is increased by 10% and 20%. What is the total percent of the population is increased after two years?

   A. 30%

   B. 32%

   C. 34%

   D. 68%

28) A basket contains 20 balls and the average weight of each of these balls is 25 g. The five heaviest balls have an average weight of 40 g each. If we remove the five heaviest balls from the basket, what is the average weight of the remaining balls.

29) How many tiles of 4 cm^2 are needed to cover a floor of dimension 5 cm by 20 cm?

   A. 5

   B. 20

   C. 25

   D. 35

30) If $a = 120°$ and $b = 98°$, what is the value of angle c?(Figure not drawn to scale.)

   A. 19°

   B. 22°

   C. 35°

   D. 45°

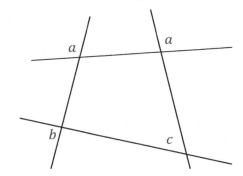

31) What is the slope of a line that is perpendicular to the line $3x - y = 6$?

A. -3

B. $-\frac{1}{3}$

C. 2

D. 6

32) The sum of 8 numbers is greater than 240 and less than 320. Which of the following could be the average (arithmetic mean) of these numbers?

A. 30

B. 35

C. 40

D. 45

33) The mean of 40 test scores was initially calculated as 60. However, it was later discovered that one of the scores was misread as 94 instead of 74. What is the corrected mean of the test scores?

A. 57

B. 59.5

C. 60.5

D. 62

34) The length of a rectangle is 3 meters greater than 4 times its width. The perimeter of the rectangle is 36 meters. What is the area of the rectangle in square meters?

   A. 35

   B. 45

   C. 55

   D. 65

35) Emily lives $5\frac{1}{4}$ miles from where she works. When traveling to work, she walks to a bus stop $\frac{1}{3}$ of the way to catch a bus. How many miles away from her house is the bus stop?

   A. $4\frac{1}{3}$ miles

   B. $4\frac{3}{4}$ miles

   C. $2\frac{3}{4}$ miles

   D. $1\frac{3}{4}$ miles

36) A bread recipe requires $2\frac{2}{3}$ cups of flour. If you only have $1\frac{5}{6}$ cups available, how much more flour do you need?

   A. 1

   B. $\frac{1}{2}$

   C. 2

   D. $\frac{5}{6}$

37) The perimeter of the trapezoid given below is 40 *cm*. What is its area?

Write your answer in the box below.

38) In 1999, the average worker's income increased $3,000 per year, starting from a $24,000 annual salary. Which equation represents income that is greater than the average? (I = income, x = number of years after 1999)

A. $I > 3,000x + 24,000$

B. $I > -3,000x + 24,000$

C. $I < -3,000x + 24,000$

D. $I < 3,000x - 24,000$

39) Which of the following graphs represents the compound inequality $-1 \leq 2x - 3 < 1$?

A.
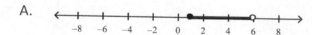

B.
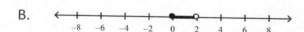

C.

D.

40) If x is directly proportional to the square of y, and $y = 2$ when $x = 12$, then when $x = 75$,

$y =$?

A. $\frac{1}{5}$

B. 1

C. 5

D. 12

End of STAAR Grade 7 Math Practice Test 7

STAAR Mathematics

Practice Test 8

2024

Grade 7

Total number of questions: 40

Total time to complete the test: No time limit

You may NOT use a calculator.

105

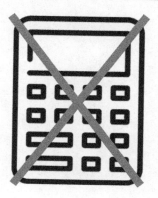

STAAR Grade 7 Mathematics Formula Sheet

LINEAR EQUATIONS

Slope – intercept form	$y = mx + b$
Direct Variation	$y = kx$
Slope of a Line	$m = \dfrac{y_2 - y_1}{x_2 - x_1}$

CIRCUMFERENCE

Circle	$C = 2\pi r$ or $C = \pi d$

AREA

Triangle	$A = \dfrac{1}{2}bh$
Parallelogram	$A = bh$
Trapezoid	$A = \dfrac{1}{2}h(b_1 + b_2)$
Circle	$A = \pi r^2$

SURFACE AREA

	Lateral	Total
Prism	$S = Ph$	$S = Ph + 2B$
Cylinder	$S = 2\pi rh$	$S = 2\pi rh + 2\pi r^2$

VOLUME

Prism or Cylinder	$V = Bh$
Pyramid or Cone	$V = \dfrac{1}{3}Bh$
Sphere	$V = \dfrac{4}{3}\pi r^3$

ADDITIONAL INFORMATION

Pythagorean theorem	$a^2 + b^2 = c^2$
Simple interest	$I = prt$
Compound Interest	$A = p(1 + r)^t$

1) What is the area of the shaded region?

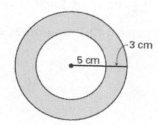

   A. $9\pi\ cm^2$

   B. $25\pi\ cm^2$

   C. $39\pi\ cm^2$

   D. $64\pi\ cm^2$

2) A pizza was cut into 8 parts. William and his sister Ella ordered two pizzas. William ate $\frac{1}{4}$ of his pizza and Ella ate $\frac{1}{2}$ of her pizza. What part of the two pizzas was left?

   A. $\frac{1}{2}$

   B. $\frac{1}{3}$

   C. $\frac{3}{8}$

   D. $\frac{5}{8}$

3) Simplify $\frac{48x^3y^8}{8x^2y^5}$?

   A. $8x^4y^6$

   B. $8x^8y^6$

   C. $6xy^3$

   D. $6x^8y^6$

4) The following table represents the value of x and function $f(x)$. Which of the following could be the equation of the function $f(x)$?

   A. $f(x) = x^2 - 5$

   B. $f(x) = x^2 - 1$

   C. $f(x) = \sqrt{x + 2}$

   D. $f(x) = \sqrt{x} + 4$

x	$f(x)$
1	5
4	6
9	7
16	8

5) Anna bought 12.9 gallons of juice. There are approximately 3.8 liters in 1 gallon. Which measurement is closest to the number of liters of juice Anna bought?

A. 24.6 L

B. 49.02 L

C. 52.08 L

D. 142 L

6) The measure of the angles of a triangle are in the ratio $1:3:5$. What is the measure of the largest angle?

A. $20°$

B. $45°$

C. $85°$

D. $100°$

7) Triangle ABC and triangle EFG are similar. Which proportion can be used to find the length of FG?

A. $\frac{8}{12} = \frac{FG}{FG}$

B. $\frac{12}{14.5} = \frac{FG}{8}$

C. $\frac{14.5}{12} = \frac{8}{FG}$

D. $\frac{12}{8} = \frac{8}{FG}$

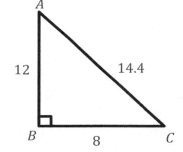 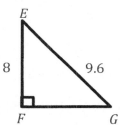

8) The price of a bar of chocolate was increased from $28.00 to $35.00. By what percentage was the price of the chocolate bar increased?

A. 18%

B. 25%

C. 27%

D. 35%

9) A store offered a 15% discount off the regular price of a desk. The amount of the discount is $6. What is the regular price of the desk?

   A. $35

   B. $40

   C. $45

   D. $50

10) The area of a triangle is 36 square inches. The base of the triangle is 4 inches. What is the height of the triangle in inches?

   A. $14\ in$

   B. $16\ in$

   C. $18\ in$

   D. $20\ in$

11) Which equation represents the relationship between x and y in the graph?

   A. $y = x + 2$

   B. $y = 2x - 1$

   C. $y = x - 2$

   D. $y = \frac{1}{2}x + 3$

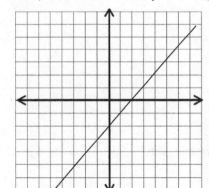

12) What is the value of the expression -8×3.2?

   Write your answer in the box below.

13) The following trapezoids are similar. What is the value of x?

A. 7

B. 8

C. 18

D. 45

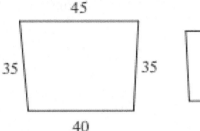 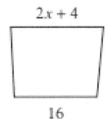

14) If $x = -8$, which equation is true?

A. $x(2x - 4) = 120$

B. $8(4 - x) = 96$

C. $2(4x + 6) = 79$

D. $6x - 2 = -46$

15) In a bag of small balls, $\frac{1}{3}$ are black, $\frac{1}{6}$ are white, $\frac{1}{4}$ are red, and the remaining 12 are blue.

How many balls are white?

A. 8

B. 12

C. 16

D. 24

16) Which values from the set $\{-5, -1, 1, 2, 5, 8\}$ satisfy this inequality? $-3x + 2 > 4$

A. $-5, 2, 8$

B. $-5, -1$

C. -5 only

D. $1, 2, 5, 8$

17) $\frac{3}{8}$ of the people who attended a football game arrived late. What percentage is equivalent

to the fraction of people who arrived late?

A. 45%

B. 15%

C. 20%

D. 37.5%

18) In the figure below, what is the value of x?

A. 43°

B. 67°

C. 77°

D. 90°

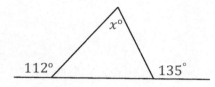

19) A wright uses 4,400 cm^2 of wood to make 40 windows of the same size. At this rate, how many square centimeters of wood are needed to make 1 window?

A. 40 cm^2

B. 95 cm^2

C. 100 cm^2

D. 110 cm^2

20) William is buying gifts for his students. He will pay $12.50 for each gift plus a one–time fee of $20 for packing. Which equation can be used to find y, the total cost to buy x gifts?

A. $y = 20x + 12.5$

B. $y = 12.5x + 20$

C. $y = 125x + 20$

D. $y = 0.125x + 20$

21) What is the value of the following expression? $|-5| + 9 \times 2\frac{1}{3} + (-3)^2$

A. 26

B. 35

C. 43

D. 51

22) The ratio of men to women in a library is 1 to 3. There are 24 women in the library. How many men are in the library?

A. 3

B. 8

C. 11

D. 15

23) What is the range of these numbers? 11, 2, 6, 15, 22, 8

   A. 6

   B. 8

   C. 15

   D. 20

24) What is the value of x in the given figure below?

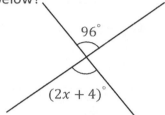

   A. 32°

   B. 46°

   C. 54°

   D. 63°

25) What is the perimeter of a square that has an area of 595.36 square feet?

   Write your answer in the box below.

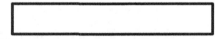

26) How many different two-digit numbers can be formed from the digits 6, 7, and 5, assuming the numbers must be even and no digit can be repeated?

   A. 1

   B. 2

   C. 3

   D. 4

27) The distance between two cities is 80 kilometers. There are approximately 16 kilometers in 10 miles. What is the distance between the two cities in miles?

   A. 10 miles

   B. 26 miles

   C. 34 miles

   D. 50 miles

28) Jason needs a 75% average in his writing class to pass. On his first 4 exams, he earned scores of 68%, 72%, 85%, and 90%. What is the minimum score Jason can earn on his fifth and final test to pass?

Write your answer in the box below.

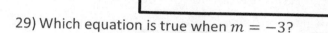

29) Which equation is true when $m = -3$?

A. $m + 10 = 5$

B. $\frac{m+6}{-3} = -1$

C. $\frac{2m+1}{2} = 3$

D. $2m - 8 = 3$

30) Jennifer has a collection of 1,100 stamps. Of these stamps, 10% are dated before 1900, 60% are dated from 1900 to 2002, and the rest are dated after 2002. How many of Jennifer's stamps are dated after 2002?

A. 112

B. 221

C. 330

D. 410

31) Jack earns $616 for his first 44 hours of work in a week and is then paid 1.5 times his regular hourly rate for any additional hours. This week, Jack needs $826 to cover his rent, bills and other expenses. How many hours must he work to make enough money this week?

A. 40

B. 48

C. 54

D. 58

32) The price of a laptop is $648 plus 5% sales tax. What is the amount of sales tax on this laptop in dollars and cents?

Write your answer in the box below.

33) An equation is represented in a model. What is the solution for this equation?

A. $x = 1$

B. $x = 3$

C. $x = 5$

D. $x = 9$

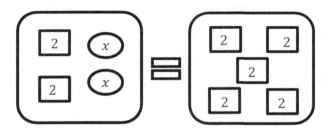

34) Which of the following could be the value of x if $\frac{6}{8} + x > 2$?

A. $\frac{1}{3}$

B. $\frac{3}{5}$

C. $\frac{6}{5}$

D. $\frac{4}{3}$

35) What is the value of x in the following figure? (The figure is not drawn to scale)

A. 15

B. 22

C. 26

D. 41

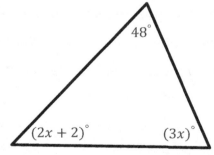

36) You can buy 5 cans of green beans at a supermarket for $3.40. How much does it cost to buy 35 cans of green beans?

A. $17.00

B. $23.80

C. $34.00

D. $119

37) Each number in a sequence is 4 more than twice the number that comes just before it. If 84 is a number in the sequence, what number comes directly before it?

A. 26

B. 35

C. 40

D. 52

38) Joe scored 20 out of 25 marks in Algebra, 30 out of 40 marks in Science and 68 out of 80 marks in Mathematics. In which subject did he achieve the best percentage of marks?

A. Algebra

B. Science

C. Mathematics

D. Algebra and Science

39) What is the volume of the following square pyramid?

A. $120 \ m^3$

B. $144 \ m^3$

C. $480 \ m^3$

D. $1,440 \ m^3$

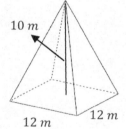

10 m

12 m

12 m

40) Robert is preparing to run a marathon. He runs $7\frac{1}{5}$ miles on Saturday and two times that many on Monday and Wednesday. Robert wants to run a total of 60 miles this week. How many more miles does he need to run?

Write your answer in the box below.

End of STAAR Grade 7 Math Practice Test 8

STAAR Mathematics

Practice Test 9

2024

Grade 7

Total number of questions: 40

Total time to complete the test: No time limit

You may NOT use a calculator.

117

STAAR Grade 7 Mathematics Formula Sheet

LINEAR EQUATIONS	
Slope – intercept form	$y = mx + b$
Direct Variation	$y = kx$
Slope of a Line	$m = \dfrac{y_2 - y_1}{x_2 - x_1}$

CIRCUMFERENCE	
Circle	$C = 2\pi r$ or $C = \pi d$

AREA	
Triangle	$A = \dfrac{1}{2}bh$
Parallelogram	$A = bh$
Trapezoid	$A = \dfrac{1}{2}h(b_1 + b_2)$
Circle	$A = \pi r^2$

SURFACE AREA		
	Lateral	Total
Prism	$S = Ph$	$S = Ph + 2B$
Cylinder	$S = 2\pi rh$	$S = 2\pi rh + 2\pi r^2$

VOLUME	
Prism or Cylinder	$V = Bh$
Pyramid or Cone	$V = \dfrac{1}{3}Bh$
Sphere	$V = \dfrac{4}{3}\pi r^3$

ADDITIONAL INFORMATION	
Pythagorean theorem	$a^2 + b^2 = c^2$
Simple interest	$I = prt$
Compound Interest	$A = p(1 + r)^t$

1) What is the slope of a line that is parallel to the line with equation of $2x - y = 12$?

   A. -2

   B. 2

   C. 4

   D. 12

2) Ryan needs $6\frac{1}{2}$ meters of wood to make a desk. He has one piece of wood that is $1\frac{1}{6}$ meters and another piece of wood that is $3\frac{1}{2}$ meters. How many more meters of wood does Ryan need to make the desk?

   A. $\frac{5}{6}$

   B. $2\frac{4}{5}$

   C. $1\frac{5}{6}$

   D. $3\frac{1}{6}$

3) A pair of shoes originally priced at $45.00 was on sale for 15% off. Nick received a 20% employee discount applied to the sale price. How much did Nick pay for the shoes?

   A. $30.60

   B. $34.50

   C. $37.30

   D. $42.25

4) Mia has answered $\frac{42}{48}$ of the questions on a mathematics test. What percentage of the questions has Mia not answered?

   A. 6%

   B. 12.5%

   C. 48.6%

   D. 87.5%

5) In five successive hours, a car travels 40 km, 45 km, 50 km, 35 km and 55 km. In the next five hours, it travels with an average speed of 50km per hour. Find the total distance the car traveled in 10 hours.

   A. 425 km

   B. 450 km

   C. 475 km

   D. 500 km

6) Nick paid $57.00 for 6 DVD. Each DVD cost the same amount. What was the cost of each DVD?

   Write your answer in the box below.

7) In box of blue and black marbles, the ratio of blue marbles to black marbles is 4 : 3. If the box contains 150 black marbles, how many blue marbles are there?

   Write your answer in the box below.

8) The list below shows the number of miles that John drives each week. What is the difference between the mode and median of this data?

   A. 22

   B. 35

   C. 54

   D. 60

Number of miles
35
54
30
76
60
38
76

9) From the figure, which of the following must be true? (Figure not drawn to scale)

A. $y = z$

B. $y = 5x$

C. $y + 4x = z$

D. $4y + x = z$

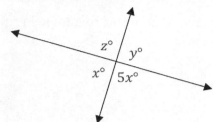

10) A football team had $20,000 to spend on supplies. The team spent $14,000 on new balls. New sport shoes cost $120 each. Which of the following inequalities represent how many new shoes the team can purchase?

A. $120x + 14,000 \leq 20,000$

B. $120x + 14,000 \geq 20,000$

C. $14,000x + 12,0 \leq 20,000$

D. $14,000x + 12,0 \geq 20,000$

11) Which expression is represented by the model below?

A. $4.\,(8)$

B. $4.\,(4)$

C. $-4.\,(4)$

D. $(-4).\,(-4)$

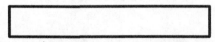

12) If $4n - 3 \geq 1$, what is the least possible value of $4n + 3$?

Write your answer in the box below.

13) The height of a scale model of a building in which 3 inches represents 45 feet is 1.5 feet. What is the height of the building?

A. $270\ ft$

B. $300\ ft$

C. $330\ ft$

D. $480\ ft$

14) The width of a rectangle is 1 meter and its length is 2 meters. Which equation can be used to determine A, the area of this rectangle in square centimeters?

   A. $\frac{2}{100} cm^2$

   B. $100(200) cm^2$

   C. $\frac{200\times100}{10} cm^2$

   D. $100(2)\ cm^2$

15) Which of the following is equivalent to $13 < -3x - 2 < 22$

   A. $-8 < x < -5$

   B. $5 < x < 8$

   C. $\frac{11}{3} < x < \frac{20}{3}$

   D. $\frac{-20}{3} < x < \frac{-11}{3}$

16) Richard has a rectangular garden. The garden is similar to a rectangle that measures 6 meters by 8 meters. What could be the dimensions of Richard's garden?

   A. $12\ m$ by $10\ m$

   B. $8\ m$ by $16\ m$

   C. $6\ m$ by $8\ m$

   D. $12\ m$ by $10\ m$

17) Figure ABC and figure $A'B'C'$ are shown on the coordinate grid below. Which statement describes how figure ABC was transformed to form the image $A'B'C'$?

   A. A reflection across the x −axis

   B. A reflection across the origin

   C. A translation across the x −axis

   D. A reflection across the y −axis

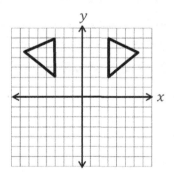

18) Jack walked 4 kilometers in 50 minutes. At this rate, how many kilometers did he walk in 20 minutes?

A. 1 km

B. 1.6 km

C. 2.5 km

D. 3.1 km

19) If $(3^a)^b = 243$, then what is the value of ab?

A. 2

B. 3

C. 4

D. 5

20) $\dfrac{1\frac{5}{4}+\frac{1}{3}}{2\frac{1}{2}-\frac{15}{8}}$ is approximately equal to ...

Write your answer in the box below.

---

21) Sarah uses 6 cups of sugar to make 12 cakes. What is the constant of proportionality that relates the number of cakes made, b, to the number of cups of sugar used, a?

A. $\dfrac{1}{3}$

B. 2

C. 6

D. 12

22) If $2x + 2y = 2$, $3x - y = 7$, which of the following ordered pairs (x, y) satisfies both equations?

A. $(1, 3)$

B. $(2, 4)$

C. $(2, -1)$

D. $(4, -6)$

23) Soap costs $3.50 per bar. Which equation best represents y, the total cost of x bars of soap?

 A. $y = 3.5x$

 B. $y = 3.5 + x$

 C. $x = 3.5y$

 D. $x = y + 3.5$

24) What is the volume of a regular square pyramid with base sides $6cm$ and height 15 cm?

 A. 81 cm^3

 B. 90 cm^3

 C. 180 cm^3

 D. 140 cm^3

25) A line in the xy −plane passes through origin and has a slope of $\frac{1}{3}$. Which of the following points lies on the line?

 A. $(2, 1)$

 B. $(4, 1)$

 C. $(9, 3)$

 D. $(6, 9)$

26) The average of five numbers is 24. If a sixth number 42 is added, then, what is the new average?

 A. 25

 B. 26

 C. 27

 D. 28

27) Emma has 2 blue marbles, 5 yellow marbles, and 3 red marbles in a bag. She will randomly choose 1 marble from the bag. What is the probability Emma will choose a yellow marble?

A. $\frac{1}{5}$

B. $\frac{5}{11}$

C. $\frac{1}{2}$

D. $\frac{1}{10}$

28) Lia created a scale drawing of her room. In the scale drawing, the width of the room is $8\ in$. The width of the actual room is $6\ m$. Which scale did she use to create the scale drawing of the room?

A. 1 inch represents $\frac{1}{8}$ meter

B. 1 inch represents $\frac{3}{4}$ meter

C. 6 inches represent $\frac{1}{8}$ meter

D. 8 inches represent $\frac{1}{6}$ meter

29) Jason deposits 15% of $160 into a savings account, what is the amount of his deposit?

A. $10

B. $16

C. $20

D. $24

30) Which of the following is equivalent to $\frac{2}{5}$?

A. 0.04

B. 0.25

C. 0.40

D. 1.4

31) Ava paid for 6 flowerpots to be delivered. Each flowerpot cost was $9.50. She paid $12.30 for delivery. What is the total amount Ava paid?

   A. $36.90

   B. $54.65

   C. $69.30

   D. $74.00

32) A tree that is 32 feet tall casts a shadow 12 feet long. Jack is 6 feet tall. How long is Jack's shadow?

   A. 2.25 feet

   B. 4 feet

   C. 4.25 feet

   D. 8.25 feet

33) What is the area of the shaded region?

   A. 31 ft^2

   B. 40 ft^2

   C. 64 ft^2

   D. 80 ft^2

34) Kim paid $30.40 for 3.2 pounds of meat. What is the price per pound of the meat?

   A. $4.30

   B. $9.50

   C. $11.54

   D. $23.40

35) Five years ago, Amy was three times as old as Mike was then. If Mike is 10 years old now, how old is Amy?

   A. 4

   B. 8

   C. 12

   D. 20

36) The circumference of a circle is A inches. The radius of the circle is 9 inches. Which expression best represents the value of π?

A. $\dfrac{A}{9}$

B. $9A$

C. $18A$

D. $\dfrac{A}{18}$

37) What is the product of all possible values of x in the following equation?

$$|x - 10| = 4$$

A. 3

B. 7

C. 13

D. 84

38) The sales price of a laptop is $1,912.50, which is 15% off the original price. What is the original price of the laptop?

Write your answer in the box below.

39) The area of a circle is 36π. What is the diameter of the circle?

A. 4

B. 8

C. 12

D. 16

40) If Sam spent $60 on sweets and he gave 25% of the selling price for the tip, how much did he spend in total?

A. $66

B. $69

C. $72

D. $75

End of STAAR Grade 7 Math Practice Test 9

STAAR Mathematics

Practice Test 10

2024

Grade 7

Total number of questions: 40

Total time to complete the test: No time limit

You may NOT use a calculator.

129

STAAR Grade 7 Mathematics Formula Sheet

LINEAR EQUATIONS

Slope – intercept form	$y = mx + b$
Direct Variation	$y = kx$
Slope of a Line	$m = \dfrac{y_2 - y_1}{x_2 - x_1}$

CIRCUMFERENCE

Circle	$C = 2\pi r \text{ or } C = \pi d$

AREA

Triangle	$A = \dfrac{1}{2}bh$
Parallelogram	$A = bh$
Trapezoid	$A = \dfrac{1}{2}h(b_1 + b_2)$
Circle	$A = \pi r^2$

SURFACE AREA

	Lateral	Total
Prism	$S = Ph$	$S = Ph + 2B$
Cylinder	$S = 2\pi rh$	$S = 2\pi rh + 2\pi r^2$

VOLUME

Prism or Cylinder	$V = Bh$
Pyramid or Cone	$V = \dfrac{1}{3}Bh$
Sphere	$V = \dfrac{4}{3}\pi r^3$

ADDITIONAL INFORMATION

Pythagorean theorem	$a^2 + b^2 = c^2$
Simple interest	$I = prt$
Compound Interest	$A = p(1 + r)^t$

1) If $y = 5ab + 3b^3$, what is y when $a = 2$ and $b = 3$?

   A. 24

   B. 31

   C. 51

   D. 111

2) To buy a car, Nina borrowed $5,000 at 6% simple interest for 3 years. How much money will she have to pay back?

   A. $900

   B. $5,500

   C. $5,650

   D. $5,900

3) Rectangle $ABCD$ is shown on the grid below. If rectangle $ABCD$ is reflected across the x −axis to form rectangle $A'B'C'D'$, which ordered pair represents the coordinates of B'?

   A. $(4, 1)$

   B. $(1, -4)$

   C. $(-1, 4)$

   D. $(-1, -4)$

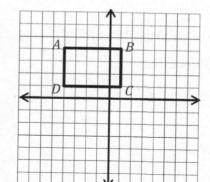

4) Yesterday Kylie wrote 10% of her homework. Today she wrote another 18% of the total homework. What fraction of the homework is left for her to write?

   Write your answer in the box below.

5) A taxi driver earns $9 per hour of work. If he works 10 hours a day and uses 2 −liters of petrol per hour at a price of $1 for 1-liter, how much money does he earn in one day?

   A. $90

   B. $88

   C. $70

   D. $60

6) If 75% of a class is girls, and $\frac{1}{3}$ of the girls take a drawing class this semester, what percent of the class is girls who take a drawing class this semester?

   A. 25%

   B. 28%

   C. 35%

   D. 37.5%

7) A barista averages making 18 coffees per hour. At this rate, how many hours will it take until she's made 1,800 coffees?

   A. 95 hours

   B. 90 hours

   C. 100 hours

   D. 105 hours

8) The dimensions of a triangular prism are shown in the diagram. What is the volume of the triangular prism in cubic meters?

   A. $18\ m^3$

   B. $24.6\ m^3$

   C. $44\ m^3$

   D. $52.5\ m^3$

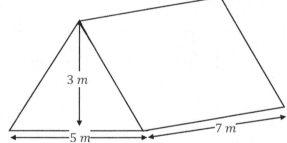

9) The sum of the measures of angle A and angle B is $180°$. If the measure of angle A is $(2x + 6)°$ and the measure of angle B is $84°$. What is the value of x?

Write your answer in the box below.

10) A gym club has 28 members. Each member pays weekly dues of $5. On the first day of the week, 6 members paid their dues. The remaining members paid their dues on the second day of the week. How much money was collected in dues on the second day of the week?

A. $30

B. $68

C. $81

D. $110

11) If $8 < x \leq 10$, then x cannot be equal to:

A. 8

B. 9

C. 9.5

D. 10

12) Use the diagram provided as a reference. If the length between point A and C is 68, and the length between point A and B is 25, what is the length between point B and C?

A. 31

B. 38

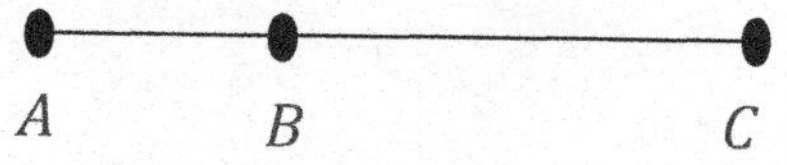

C. 41

D. 43

A B C

13) Clara is 4 feet 8 inches tall. There are 25.4 millimeters in 1 inch. What is Clara's height in millimeters?

   A. 1,000.3 mm

   B. 1,036.5 mm

   C. 1,422.4 mm

   D. 1,850.3 mm

14) If the area of the following rectangular $ABCD$ is 160, and E is the midpoint of AB, what is the area of the shaded part?

   Write your answer in the box below.

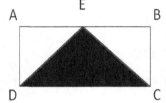

15) If $8a = -4$, which inequality is true?

   A. $1 - 6a < 2$

   B. $-2 + 4a > 2$

   C. $-8a + 1 < 0$

   D. $-10a - 1 > 3$

16) The volume of a cube is less than $64\ m^3$. Which of the following can be the cube's side?

   A. $2\ m$

   B. $4\ m$

   C. $5\ m$

   D. $8\ m$

17) If 60% of A is 20% of B, then B is what percent of A?

   A. 3%

   B. 30%

   C. 200%

   D. 300%

18) If Ella needed to buy 40 bottles of soda for a party in which 100 people attended, how many bottles of soda will she need to buy for a party in which 5 people are attending?

   Write your answer in the box below.

19) A square has an area of 121 cm^2. What is its perimeter?

   A. 28 cm

   B. 36 cm

   C. 38 cm

   D. 44 cm

20) A car traveled 42 miles in 3 hours. At this rate, how many miles will the car travel in $\frac{1}{7}$ hour?

   A. 2 miles

   B. 7 miles

   C. $2\frac{1}{7}$ miles

   D. $3\frac{1}{6}$ miles

21) If $2 \leq x < 6$, what is the minimum value of the following expression?

$$3x + 1$$

A. 9

B. 7

C. 4

D. 2

22) The route that John takes from his house to his school is represented on the grid below. Which ordered pair represents a point on John's route?

A. $(3, 1)$

B. $(1, 4)$

C. $(-2, 3)$

D. $(-4, -2)$

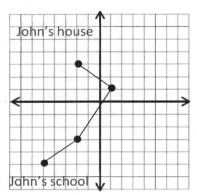

23) A number is chosen at random from 1 to 20. Find the probability of not selecting a composite number. (A composite number is a number that is divisible by itself, 1 and at least one other whole number)

A. $\frac{1}{20}$

B. $\frac{2}{5}$

C. $\frac{9}{20}$

D. 1

24) If Logan ran 1.25 miles in 15 minutes, what was his average speed?

    A. 1.25 miles per hour

    B. 2.5 miles per hour

    C. 3.75 miles per hour

    D. 5 miles per hour

25) A circle has a diameter of 12 centimeters. Which measurement is closest to the circumference of the circle in centimeters? ($\pi = 3.14$)

    A. 37.7 cm

    B. 41.36 cm

    C. 50.77 cm

    D. 61.5 cm

26) What is the cube root of 3,375?

    A. 155

    B. 15

    C. 7.5

    D. 5

27) Rectangle A has a length of 10 cm and a width of 6 cm, and rectangle B has a length of 6 cm and a width of 4 cm. What is the ratio (expressed as a percentage) of the perimeter of rectangle B to rectangle A?

    A. 10.25%

    B. 20%

    C. 62.5%

    D. 75%

28) $\frac{1}{6b^2} + \frac{1}{6b} = \frac{1}{b^2}$, then $b =$?

    A. $-\frac{16}{15}$

    B. 5

    C. $-\frac{15}{16}$

    D. 8

29) The sum of two numbers is N. If one of the numbers is 6, what is three times the other number?

    A. $3N$

    B. $3(N - 6)$

    C. $3(N + 6)$

    D. $(N - 2)$

30) There are three equal tanks of water. If $\frac{2}{5}$ of a tank contains 200 liters of water, what is the capacity of the three tanks of water together?

    A. 1,500

    B. 500

    C. 240

    D. 80

31) Which of the following shows the numbers in decreasing order?

    A. $\frac{1}{5}, \frac{5}{3}, \frac{8}{11}, \frac{2}{3}$

    B. $\frac{5}{3}, \frac{8}{11}, \frac{2}{3}, \frac{1}{5}$

    C. $\frac{8}{11}, \frac{2}{3}, \frac{5}{3}, \frac{1}{5}$

    D. $\frac{2}{3}, \frac{8}{11}, \frac{5}{3}, \frac{1}{5}$

32) The ratio of two sides of a parallelogram is $2:3$. If its perimeter is $40\ cm$, what are the lengths of its sides?

A. $8\ cm, 12\ cm$

B. $10\ cm, 14\ cm$

C. $12\ cm, 16\ cm$

D. $14\ cm, 18\ cm$

33) Which arithmetic sequence is represented by the expression $5m - 1$, where m represents the position of a term in the sequence?

A. $3, 8, 15, 19, ...$

B. $5, 11, 19, 24, ...$

C. $4, 9, 14, 19, ...$

D. $4, 7, 13, 18, ...$

34) The following list shows Emma's scores in her last six tests. Which measure of data best describes the variation in these scores? $65, 98, 81, 57, 40, 60$

A. Range

B. Mean

C. Mode

D. Median

35) Alice is choosing a menu for her lunch. She has 3 choices of appetizers, 5 choices of entrees, and 6 choices of cake. How many different menu combinations can she choose from?

A. 12

B. 32

C. 90

D. 120

36) The Jackson Library is ordering some bookshelves. If x is the number of bookshelves the library wants to order, and each costs $200 with a one-time delivery charge of $600, which of the following represents the total cost, in dollar, per bookshelf?

A. $\dfrac{200x+600}{x}$

B. $\dfrac{200x+600}{200}$

C. $200 + 600x$

D. $200x + 600$

37) Sara graphed a square on the grid below. Sara translated the square 3 units to the right and 5 units down. Which grid shows the result of this translation?

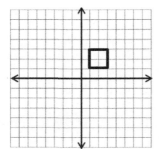

A.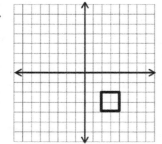

B.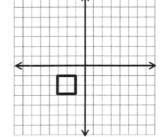

C.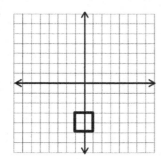

D.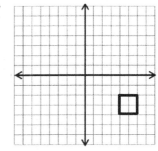

38) The ratio of $5a$ to $2b$ is $\frac{1}{10}$, what is the ratio of a to b?

    A. 10

    B. 25

    C. $\frac{1}{25}$

    D. $\frac{1}{20}$

39) If x can be any integer, what is the greatest possible value of the expression $2 - x^2$?

    A. -1

    B. 0

    C. 3

    D. 2

40) Of 120 balls in a bag, 25% are red. What is the total number of red balls in the bag?

    A. 24

    B. 30

    C. 36

    D. 41

End of STAAR Grade 7 Math Practice Test 10

STAAR Mathematics Practice Tests Answer Keys

Now, it's time to review your results to see where you went wrong and what areas you need to improve.

STAAR Math Practice Test 1				STAAR Math Practice Test 2			
1	A	21	44	1	B	21	B
2	A	22	B	2	A	22	A
3	A	23	C	3	D	23	A
4	B	24	A	4	B	24	A
5	C	25	A	5	$\frac{7}{12}$	25	A
6	D	26	B	6	C	26	C
7	A	27	D	7	D	27	C
8	D	28	D	8	D	28	B
9	B	29	C	9	A	29	C
10	−3.40	30	B	10	D	30	B
11	C	31	D	11	90	31	C
12	A	32	D	12	B	32	B
13	C	33	28	13	B	33	D
14	D	34	C	14	A	34	C
15	C	35	A	15	400	35	A
16	D	36	B	16	A	36	A
17	A	37	D	17	C	37	B
18	C	38	A	18	B	38	C
19	B	39	12	19	$\frac{64}{343}$	39	A
20	A	40	A	20	C	40	D

STAAR Math Practice Test 3				STAAR Math Practice Test 4			
1	C	21	D	1	B	21	D
2	A	22	C	2	D	22	C
3	B	23	49	3	B	23	C
4	B	24	A	4	A	24	B
5	D	25	B	5	A	25	D
6	D	26	C	6	C	26	30
7	C	27	A	7	A	27	A
8	B	28	A	8	C	28	D
9	C	29	B	9	90	29	25
10	314	30	B	10	B	30	A
11	D	31	D	11	B	31	B
12	C	32	A	12	A	32	A
13	A	33	D	13	D	33	D
14	B	34	A	14	A	34	C
15	D	35	12	15	C	35	C
16	54°	36	B	16	B	36	A
17	A	37	A	17	A	37	A
18	D	38	D	18	D	38	$\frac{1}{4}$
19	C	39	B	19	D	39	B
20	A	40	A	20	A	40	C

STAAR Math Practice Test 5				STAAR Math Practice Test 6			
1	A	21	A	1	A	21	-64
2	C	22	D	2	C	22	D
3	B	23	A	3	B	23	52
4	D	24	B	4	A	24	D
5	20	25	A	5	B	25	C
6	B	26	C	6	C	26	D
7	B	27	D	7	C	27	C
8	D	28	C	8	C	28	C
9	C	29	A	9	D	29	D
10	B	30	A	10	B	30	80
11	D	31	110	11	B	31	A
12	40	32	C	12	B	32	12
13	A	33	D	13	D	33	D
14	C	34	D	14	C	34	D
15	B	35	A	15	B	35	61.2
16	A	36	B	16	D	36	C
17	A	37	A	17	D	37	D
18	B	38	C	18	D	38	B
19	D	39	D	19	D	39	175
20	$\frac{1}{3}$	40	B	20	C	40	C

STAAR Math Practice Test 7				STAAR Math Practice Test 8			
1	B	21	A	1	C	21	B
2	10	22	C	2	D	22	B
3	D	23	B	3	C	23	D
4	B	24	A	4	D	24	B
5	B	25	C	5	B	25	97.6
6	C	26	D	6	D	26	B
7	C	27	B	7	D	27	D
8	D	28	20	8	B	28	60
9	C	29	C	9	B	29	B
10	5	30	B	10	C	30	C
11	C	31	B	11	C	31	C
12	D	32	B	12	-25.6	32	32.40
13	C	33	B	13	A	33	B
14	C	34	B	14	B	34	D
15	D	35	D	15	A	35	C
16	D	36	D	16	B	36	B
17	C	37	90	17	D	37	C
18	D	38	A	18	B	38	C
19	C	39	D	19	D	39	C
20	C	40	C	20	B	40	$38\frac{2}{5}$

STAAR Math Practice Test 9				STAAR Math Practice Test 10			
1	B	21	B	1	D	21	B
2	C	22	C	2	D	22	C
3	A	23	A	3	B	23	B
4	B	24	C	4	$\frac{18}{25}$	24	D
5	C	25	C	5	C	25	A
6	9.50	26	C	6	A	26	B
7	200	27	C	7	C	27	C
8	A	28	B	8	D	28	B
9	C	29	D	9	45	29	B
10	A	30	C	10	D	30	A
11	B	31	C	11	A	31	B
12	7	32	A	12	D	32	A
13	A	33	B	13	C	33	C
14	B	34	B	14	80	34	A
15	A	35	D	15	D	35	C
16	C	36	D	16	A	36	A
17	D	37	D	17	D	37	D
18	B	38	2,250	18	2	38	C
19	D	39	C	19	D	39	D
20	$\frac{62}{15}$	40	D	20	A	40	B

How to score your test

The basic score on each STAAR test is the raw score, which is simply the number of questions correct. On the STAAR test each subject test should be passed individually. It means that you must meet the standard on each section of the test. If you failed one subject test but did well enough on another, that's still not a passing score.

There are four possible scores that you can receive on the STAAR Math Grade 7 Test:

Do Not Meet: This indicates that your score is lower than the passing score. If you do not pass, you can reschedule to retake any the STAAR Math test. Students have three opportunities to retake test(s) and receive remedial help if they don't pass.

Approaches: This level was previously known as Satisfactory. This score indicates that your score meets the standard of the test. A student achieving Approaches Grade Level is likely to succeed in the next grade or course with targeted academic intervention.

Met the Standard: This indicates that your score meets Texas state standards for that subject. Students in this category generally demonstrate the ability to think critically and apply the assessed knowledge and skills in familiar contexts.

Commended Performance: This indicates that you've mastered the skills that would be taught in your grade.

There are approximately 40 questions on STAAR Mathematics for grade 7. Similar to other subject areas, you will need a minimum score to pass the Mathematics Test. There are approximately 40 raw score points on the STAAR math test. The raw points correspond with correct answers. This will then be converted into your scaled score. Approximately, you need to get 28 out of 40 raw score to pass the STAAR Mathematics for grade 7.

To score your STAAR Mathematics practice tests, first find your raw score. There were 40 questions on each STAAR Mathematics practice test in this book. All questions have one point. Use the following table to convert your raw score to the scale score.

Raw Score	Scale Score	Result	Percentile
0	1065		0
1	1198		0
2	1277		0
3	1326		0
4	1362		1
5	1391		1
6	1416		2
7	1437		4
8	1457	Do Not Meet	8
9	1475		12
10	1491		17
11	1507		23
12	1522		39
13	1536		34
14	1550		40
15	1563		44
16	1575		50
17	1589		53
18	1602		57
19	1615		60
20	1627	Approaches	63
21	1640		67
22	1653		70
23	1665		72
24	1678		75
25	1688		77
26	1706		80
27	1720		82
28	1734	Meets	84
29	1750		86
30	1766		88
31	1784		90
32	1798		91
33	1823		93
34	1846		95
35	1872		96
36	1903	Masters	97
37	1941		98
38	1993		99
39	12076		100
40	2212		100

STAAR Mathematics Practice Tests Answers and Explanations

STAAR Mathematics Practice Test 1
Answers and Explanations

1)   Choice A is correct.

To determine the probability of a randomly selected student's favorite subject, you need to divide the number of students who chose each subject by the total number of students who participated in the survey.

The total number of students who participated in the survey is $30 + 20 + 40 + 10 = 100$.

The probability of a randomly selected student's favorite subject being Math is $\frac{30}{100} = 0.3$, Science is $\frac{20}{100} = 0.2$, English is $\frac{40}{100} = 0.4$, and History is $\frac{10}{100} = 0.1$.

Therefore, English has the highest probability of being the favorite subject, making option A the correct answer. Option B is incorrect as the probability of a student's favorite subject being Math is not twice as likely as Science. Option C is incorrect as the probabilities of each subject are not equal. Option D is incorrect as the probability of a student's favorite subject being History is not more than twice as likely as Science.

2)   Choice A is correct.

To convert the length of the rope from feet to meters, we need to multiply the length in feet by the conversion factor of $0.3048 \frac{meters}{foot}$.

Length of rope in meters $= 15 \, feet \times 0.3048 \frac{meters}{foot} = 4.572 \, meters$

Therefore, the measurement closest to the length of the rope in meters is A.

3)   Choice A is correct.

To find the value of y, we need to use the fact that angle A and angle B are complementary, which means their sum is 90 degrees.

Therefore, the equation that can be used to find the value of y is:

A. $35 + (6y - 11) = 90$

4)   Choice B is correct.

To find the sales tax on the book, we need to multiply the price of the book by the sales tax rate, which is expressed as a decimal.

The sales tax rate of 6% can be written as 0.06.

$Sales\ tax = Price\ of\ the\ book \times Sales\ tax\ rate$

$Sales\ tax = \$13 \times 0.06$

$Sales\ tax = \$0.78$

Therefore, the sales tax on this book is $0.78.

5)   Choice C is correct.

To solve this problem, we can use the formula: $distance = rate \times time$

where distance is the distance traveled, rate is the speed of the car, and time is the time traveled.

We are given that the car traveled 84 miles in 6 hours. So, the rate of the car is: $rate = \frac{distance}{time}$

$rate = \frac{84\ miles}{6\ hours} = 14 \frac{miles}{hour}$

Now, we need to find out how many miles the car will travel in $\frac{1}{3}$ hour. Using the formula above,

we get: $distance = rate \times time$

$distance = 14 \frac{miles}{hour} \times \left(\frac{1}{3}\right) hour = \frac{14}{3} miles$

So, the car will travel $4 \frac{2}{3}$ miles in $\frac{1}{3}$ hour.

6)   Choice D is correct.

Divide the shape into trapezoid and triangle, and label the shapes as A and B. Then, calculate the area of each shape. The area of trapezoid A is $\left(\frac{18+8}{2}\right) \times 10 = 130\ m^2$, and the area of triangle B

is $\frac{1}{2}(19 - 10) \times 18 = 81\ m^2$

Summing the area of shapes, A and B yields $130 + 81 = 211\ m^2$.

Therefore, the total area is 211 square meters.

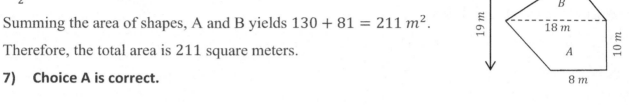

7)   Choice A is correct.

Subtract 3 from both sides of $5x + 3 \geq -2$: $5x + 3 - 3 \geq -2 - 3 \rightarrow 5x \geq -5$

Divide both sides of $5x \geq -5$ by 5: $x \geq -1$

8)   Choice D is correct.

To solve this problem, we use the formula for probability:

$$P(A \text{ and } B) = P(A) \times P(B|A)$$

where $P(A \text{ and } B)$ is the probability of both A and B occurring. $P(A)$ is the probability of A occurring, and $P(B|A)$ is the probability of B occurring given that A has occurred.

In this case, let A be the event that the first marble selected is blue, and let B be the event that the second marble selected is blue.

$P(A) = \frac{7}{22}$ (Since there are 7 blue marbles out of a total of 22 marbles)

$P(B|A) = \frac{6}{21}$ (Since there are now 6 blue marbles left out of a total of 21 marbles)

Therefore:

$$P(A \text{ and } B) = P(A) \times P(B|A) = \left(\frac{7}{22}\right) \times \left(\frac{6}{21}\right) = \frac{42}{462} = \frac{7}{77} = \frac{1}{11}$$

So, the probability of selecting two blue marbles without replacement is $\frac{1}{11}$.

Therefore, the answer is D.

9)   Choice B is correct.

Count and write each set of tiles, $6x + 2 = 3$. Solve the resulting equation to determine the value of x:

$$6x + 2 = 3 \rightarrow 6x + 2 - 2 = 3 - 2 \rightarrow 6x = 1 \rightarrow x = \frac{1}{6}$$

10) The answer is -3.40.

We can start by converting the mixed number $3\frac{3}{4}$ to an improper fraction:

$$3\frac{3}{4} = \frac{4 \times 3 + 3}{4} = \frac{15}{4}$$

Then we can divide -12.75 by $\frac{15}{4}$:

$$-12.75 \div \left(\frac{15}{4}\right) = -12.75 \times \frac{4}{15} = -3.40$$

Therefore, the answer is B.

11) Choice C is correct.

Find the corresponding sides and write a proportion.

$\frac{AB}{AD} = \frac{AC}{AE} = \frac{BC}{DE}$.

According to the size of the figure, place the sides in the ratio above:

$\frac{8}{8+4} = \frac{17}{17+8.5} = \frac{15}{DE} \rightarrow \frac{8}{12} = \frac{17}{25.5} = \frac{15}{DE}$.

Therefore, the proportion $\frac{8}{12} = \frac{15}{DE}$ is true.

12) Choice A is correct.

We can write the total cost of buying c chairs and t tables as:

$Cost = 50c + 200\,t$

Since the store has a total budget of $1,000, we can write the inequality:

$50c + 200\,t \leq 1,000$

Therefore, the inequality that represents all possible values of t and c is:

A. $200\,t + 50c \leq 1,000$.

13) Choice C is correct.

First, find the slope of the line using the formula. To use the formula, you must consider two points on the line, for example $(0, 3)$ and $(2, 7)$.

$m = \frac{y_2 - y_1}{x_2 - x_1} = \frac{7 - 3}{2 - 0} = \frac{4}{2} = 2$

The next step is to find the $y-$intercept of the line. The intersection of the line with the $y-$axis is $(0, 3)$. Therefore, the $y-$intercept is 3. The equation of the line will be as follows:

$$y = mx + b \rightarrow y = 2x + 3$$

14) Choice D is correct.

Use the following formula to calculate the surface area of a rectangular prism: $2lw + 2lh + 2hw$

Therefore, the surface area of the rectangular cube of the question is equal to:

$2lw + 2lh + 2hw = 2(5 \times 2) + 2(5 \times 14) + 2(14 \times 2) = 2 \times 10 + 2 \times 70 + 2 \times 28$

$$= 20 + 140 + 56 = 216\ cm^2$$

15) Choice C is correct.

We can use the formula for net worth:

Net worth = Assets − Liabilities

We know that Jane's net worth is $12,300, and we can add up all her assets and liabilities to find the missing value of her car:

$12,300 = Car + $3,500 + $10,000 + $45,000 − $1,200 − $120,000 − $5,000$

Simplifying the equation, we get:

$12,300 = Car − $67,700$

Adding $67,700 to both sides, we get:

$80,000 = Car$

Therefore, the current value of Jane's car is $80,000.

16) Choice D is correct.

The correct equation that represents y, the total cost of x pieces of mangoes is:

D. $y = 2.50x$

This equation shows that the total cost (y) of x pieces of mangoes is equal to the price per piece ($2.50) multiplied by the number of pieces (x). Therefore, as the number of pieces of mangoes increases, the total cost also increases proportionally.

17) Choice A is correct.

The ratio of red to white flowers in the garden is $2:5$, which means that for every 2 red flowers, there are 5 white flowers.

Let's assume that there are $2x$ red flowers and $5x$ white flowers in the garden, where x is a common factor.

According to the given information, the total number of red and white flowers in the garden is 140.

Therefore, we can write the equation: $2x + 5x = 140$

Simplifying the equation, we get: $7x = 140$

Dividing both sides by 7, we get: $x = 20$

So, there are $2x = 2 \times 20 = 40$ red flowers in the garden.

Therefore, the answer is that there are 40 red flowers in the garden.

18) Choice C is correct.

The number of people who were asked= 120

Percentage of people who like blue: = 25%

Then, the number of people who like blue: $25\% \times 120 = 0.25 \times 120 = 30$

Percentage of people who like red: = 15%

Then, the number of people who like red: $15\% \times 120 = 0.15 \times 120 = 18$

Therefore,$30 - 18 = 12$ more people like blue than red.

19) Choice B is correct.

To find the total distance Mandy traveled, we need to calculate the distance she covered while walking and running, and then add them together.

Distance covered while walking (first) $= Speed \times Time = 4\frac{km}{h} \times 1\ hour = 4\ km$

Distance covered while running $= Speed \times Time = 8\frac{km}{h} \times 2\ hours = 16\ km$

Distance covered while walking (second) $= Speed \times Time = 3\ km/h \times 1\ hour = 3\ km$

Total distance traveled = Distance covered while walking (first) + Distance covered while running + Distance covered while walking (second)

$= 4\ km + 16\ km + 3\ km = 23\ km$

Therefore, Mandy traveled a total distance of 23 km during this time.

20) Choice A is correct.

Use the following formula to calculate the volume of a rectangular pyramid:

$$v = \frac{1}{3} \times w \times l \times h$$

Therefore, the volume of the shape is equal to:

$$v = \frac{1}{3} \times w \times l \times h = \frac{1}{3} \times 8 \times 12 \times 18 = 576$$

21) The answer is 44.

In any triangle, the sum of all angles is 180 degrees, Therefore, add the internal angles of this triangle together and set it equal to 180: $(2x + 1) + (x) + (x + 3) = 180 \rightarrow$

$$4x + 4 = 180 \rightarrow 4x + 4 - 4 = 180 - 4 \rightarrow 4x = 176 \rightarrow x = 44$$

22) Choice B is correct.

The original price of the jacket is $80 and the new price is $100. To find the percentage increase, we need to calculate the difference between the new and old prices, divide that by the old price, and then multiply by 100 to get a percentage.

Difference between new and old price = $100 − $80 = $20

$$Percentage\ increase = \left(\frac{Difference}{Old\ price}\right) \times 100\% = \left(\frac{\$20}{\$80}\right) \times 100\% = 0.25 \times 100\% = 25\%$$

Therefore, the percentage increase in the price of the jacket is 25%, and the answer is B.

23) Choice C is correct.

First, find the area of each circle. Then subtract the area of the inner circle from the area of the outer circle.

The area of the inner circle, $\pi \times (5)^2 = 25\pi\ cm^2$.

The area of the outer circle, $\pi \times \left(\frac{20}{2}\right)^2 = 100\pi\ cm^2$.

To find the shaded area, $100\pi - 25\pi = 75\pi\ cm^2$.

So, the area of the shaded region is $75\pi\ cm^2$.

24) Choice A is correct.

To find out how much change John should receive, we need to first calculate the total cost of his lunch order, including sales tax:

Total cost = (Sandwich + Bag of chips + Soda) + Sales tax

Total cost = ($6.99 + $1.49 + $1.89) + $0.75

Total cost = $11.12

John paid with a $20 bill, so we need to subtract the total cost from the $20 to find the amount of change he should receive:

Change = $20 − $11.12

Change = $8.88

Therefore, John should receive $8.88 in change from his $20 bill. The answer is A.

25) Choice A is correct.

To find out which equation is true when $x = 2$, we simply substitute 2 for x in each equation and see which equation is true.

A. $2x + 4 = 8$: Substituting $x = 2$, we get: $2(2) + 4 = 8 \rightarrow 4 + 4 = 8 \rightarrow 8 = 8$

This equation is true when $x = 2$.

B. $4x - 6 = 10$: Substituting $x = 2$, we get: $4(2) - 6 = 8 - 6 \rightarrow 2 \neq 10$

This equation is false when $x = 2$.

C. $3x + 7 = 16$: Substituting $x = 2$, we get: $3(2) + 7 = 6 + 7 \rightarrow 13 \neq 16$

This equation is false when $x = 2$.

D. $5x + 3 = 14$: Substituting $x = 2$, we get: $5(2) + 3 = 10 + 3 \rightarrow 13 \neq 14$

This equation is false when $x = 2$.

Therefore, the equation that is true when $x = 2$ is A.

26) Choice B is correct.

To solve the inequality $2x + 5 > 17$, we need to isolate x on one side of the inequality sign.

Subtracting 5 from both sides gives us: $2x > 12$

Dividing both sides by 2 gives us: $x > 6$

Therefore, the solution set for the inequality $2x + 5 > 17$ is: B. $x > 6$

27) Choice D is correct.

To generate an arithmetic sequence, we start with the first term and add the common difference to each term to get the next term.

In this case, the first term is -2, and the common difference is 5.

So, the sequence can be generated as follows:

First term: -2

Second term: $-2 + 5 = 3$

Third term: $3 + 5 = 8$

Fourth term: $8 + 5 = 13$

Fifth term: $13 + 5 = 18$

...

This matches the sequence given in option D, so D is the correct answer.

28) Choice D is correct.

The measure of data that would best describe the variability of these temperature readings is D. Range.

Range is the difference between the highest and lowest values in a data set. In this case, the highest temperature is $22°C$ and the lowest temperature is $18°C$, so the range is $22°C - 18°C = 4°C$. The mean, median, and mode are measures of central tendency and do not provide information on the variability or spread of the data set.

29) Choice C is correct.

We use the volume of the triangular prism formula.

$$V = \frac{1}{2}(length)(base)(height) \rightarrow V = \frac{1}{2} \times 7 \times 5 \times 4 \rightarrow V = 70 \ m^3$$

30) Choice B is correct.

The scale ratio of the painting is 1 inch to 10 miles. So, to find the distance between the two mountains in inches, we can set up a proportion:

$$\frac{1 \ inch}{10 \ miles} = \frac{x \ inches}{70 \ miles}$$

Cross-multiplying, we get:

$10 \ miles \times x \ inches = 1 \ inch \times 70 \ miles$

Simplifying, we get: $x = 7 \ inches$

Therefore, the distance between the two mountains in the painting is 7 inches.

So, the answer is B, 7 inches.

31) Choice D is correct.

According to the chart, the interquartile range of group A data is smaller than that of group B data. Therefore, both answer A and B are not correct.

The median number of study hours in study group A is 7 and the median study hours in study group B is 5. So, the median number of study hours in study group A is greater than that of group B, making answer D the correct choice.

32) Choice D is correct.

To find the equation that represents the linear relationship between the $x-$values and the $y-$values in the given table, we can use the slope-intercept form of a linear equation: $y = mx + b$, where m is the slope and b is the $y-$intercept.

To find the slope, we can use any two points from the table and calculate the difference in $y-$values over the difference in $x-$values:

$$slope = \frac{y_2 - y_1}{x_2 - x_1}$$

Using the points $(2, 6)$ and $(3, 8)$, we get:

$$slope = \frac{8 - 6}{3 - 2} = \frac{2}{1} = 2$$

Now that we have the slope, we can use any point from the table and the slope to find the $y-$intercept. Using the point $(2,6)$, we get:

$$y = mx + b$$

$$6 = 2 \times 2 + b$$

$$6 = 4 + b \rightarrow b = 2$$

Therefore, the equation that represents the linear relationship between the $x-$values and the $y-$values in the given table is:

$$y = 2x + 2$$

33) The answer is 28.

To solve for the sum of six numbers: The average $= \frac{sum\ of\ terms}{number\ of\ terms} \rightarrow 25 = \frac{sum\ of\ 6\ numbers}{6}$

$\rightarrow sum\ of\ 6\ numbers = 25 \times 6 = 150$. The sum of 6 numbers is 150. If a seventh number 46 is added, then the sum of 7 numbers is $150 + 46 = 196$

$$The\ new\ average = \frac{sum\ of\ terms}{number\ of\ terms} = \frac{196}{7} = 28$$

34) Choice C is correct.

$1 < -2x + 5 < 15 \rightarrow$ Subtract 5 from each side. $1 - 5 < -2x + 5 - 5 < 15 - 5$

$\rightarrow -4 < -2x < 10 \rightarrow$ Divide all sides by -2. (Remember that when you divide all sides of an inequality by a negative number, the inequality sign will be swapped. $<$ becomes $>$):

$\frac{-4}{-2} > \frac{-2x}{-2} > \frac{10}{-2} \rightarrow -5 < x < 2$

35) Choice A is correct.

Find the points one by one on the coordinate plane to see which one is inside the triangle. For example, the point $(2, -1)$ is 2 units to the right of the $x-$axis and 1 unit below the $y-$axis. Find the rest of the points in the same way on the coordinate plane. After finding the points, you can see that only answer A is inside the triangle.

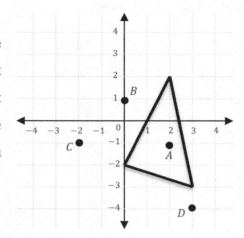

36) Choice B is correct.

The soccer team practices for a total of $4 + 3 + 2 = 9$ hours per week.

Therefore, in 3 weeks, the team will practice for a total of $3 \times 9 = 27$ hours.

So, the answer is B, 27 hours.

37) Choice D is correct.

The area of the rectangle is $6 \times 20 = 120 \ m^2$.

The area of the triangle is $\frac{1}{2}(6) \times (5) = 15 \ m^2$.

The area of the small semicircle is $\frac{1}{2} \times \pi \times (3)^2 = 14.13 \ m^2$.

The area of the large semicircle is $\frac{1}{2} \times \pi \times (4)^2 = 25.12 \ m^2$.

The total area of the figure is equal to the sum of the above areas:

$$120 + 15 + 14.13 + 25.12 = 174.25 \ m^2$$

38) Choice A is correct.

Write the numbers in order: $0, 5, 8, 12, 14, 21, 24, 27, 33$. Since we have 9 numbers (an odd quantity), then the median is the number in the middle, which is 14.

39) The answer is 12.

The cube root of 1,728 is 12, because: $1,728 = 12^3$, $\sqrt[3]{12^3} = 12$, Therefore, $\sqrt[3]{1,728} = 12$.

40) Choice A is correct.

Let x be the number. Write the equation and solve for x. $(35 - x) \div x = 4$. Multiply both sides by x. $(35 - x) = 4x$, then add x both sides. $35 = 5x$. Now divide both sides by 5 to solve for x: $x = 7$.

STAAR Mathematics Practice Test 2

Answers and Explanations

1) Choice B is correct.

The volume of a cylinder $= \pi r^2 h$, In this case, the radius of the cylinder is 3 meters and its height is 11 meters. Therefore, the volume of a cylinder $= \pi(3)^2(11) = 310.86 \ m^3$.

2) Choice A is correct.

To calculate the probability that all three balls drawn will be red, we need to find the probability of drawing a red ball on the first draw, then the probability of drawing another red ball on the second draw (since we are drawing without replacement), and then the probability of drawing a third red ball on the third draw.

The probability of drawing a red ball on the first draw is $\frac{8}{20}$, or $\frac{2}{5}$.

The probability of drawing a red ball on the second draw, given that the first ball was red, is $\frac{7}{19}$ (since there are now only 19 balls left in the bag, of which 7 are red).

The probability of drawing a red ball on the third draw, given that the first two balls were red, is $\frac{6}{18}$ (since there are now only 18 balls left in the bag, of which 6 are red).

Therefore, the probability of drawing three red balls in a row is:

$\left(\frac{8}{20}\right) \times \left(\frac{7}{19}\right) \times \left(\frac{6}{18}\right) = \frac{14}{285}$

Therefore, the answer is A.

3) Choice D is correct.

The area of the rectangle is $7 \times 14 = 98 \ m^2$

The area of the triangle is $\frac{1}{2}(7) \times (5) = 17.5 \ m^2$

The area of the small trapezoid is $\frac{1}{2} \times (14 + 16) \times 3 = 45 \ m^2$

The area of the large trapezoid is $\frac{1}{2} \times (14 + 12) \times 3 = 39 \ m^2$

The total area of the figure is equal to the sum of the above areas:

$98 + 1.5 + 45 + 39 = 199.5 \ m^2$

4)   Choice B is correct.

The formula for the volume of a square pyramid is $V = \left(\frac{1}{3}\right)Bh$, where B is the area of the base and h is the height of the pyramid.

The base of the square pyramid has sides of length 8 inches, so the area of the base is:

$B = 8^2 = 64\ in^2$

The height of the pyramid is given as 10 inches. Therefore, the volume of the pyramid is:

$V = \left(\frac{1}{3}\right)(64)(10) = 213.33\ in^3$ (Rounded to two decimal places)

Therefore, the answer is $213.33\ in^3$.

5)   The answer is $\frac{7}{12}$.

To solve the equation $\frac{3}{4}y - \frac{1}{8} = \frac{5}{16}$, we can start by getting rid of the fraction on the left-hand side. We can do this by multiplying both sides of the equation by the reciprocal of $\frac{3}{4}$, which is $\frac{4}{3}$.

$\left(\frac{4}{3}\right)\left(\frac{3}{4}y - \frac{1}{8}\right) = \left(\frac{4}{3}\right)\left(\frac{5}{16}\right) \rightarrow y - \frac{1}{6} = \frac{5}{12}$

Next, we can isolate y by adding $\frac{1}{6}$ to both sides of the equation: $y - \frac{1}{6} + \frac{1}{6} = \frac{5}{12} + \frac{1}{6} \rightarrow y = \frac{7}{12}$

Therefore, the value of y in the equation $\frac{3}{4}y - \frac{1}{8} = \frac{5}{16}$ is $\frac{7}{12}$.

6)   Choice C is correct.

Solve for x. $-5 \leq 3x - 2 < 4 \rightarrow$ (Add 2 all sides) $-5 + 2 \leq 3x - 2 + 2 < 4 + 2 \rightarrow$

$-3 \leq 3x < 6 \rightarrow$ (Divide all sides by 3) $-1 \leq x < 2$.

x is between -1 and 2. Choice C represents this inequality.

7)   Choice D is correct.

The net worth of John can be calculated by subtracting his total liabilities from his total assets.

Total Assets $= \$260,000$

Total Liabilities $= \$110,500$

Net Worth $= Total\ Assets - Total\ Liabilities$

Net Worth $= \$260,000 - \$110,500$

Net Worth $= \$149,500$

Therefore, the answer is (D) $\$149,500$.

8) Choice D is correct.

Let's review the choices when $x = 1$.

A. $f(x) = 4x^2 + 1$, if $x = 1 \rightarrow f(1) = 4(1)^2 + 1 = 4 + 1 = 5 \neq 4$

B. $f(x) = x^2 + 1$, if $x = 1 \rightarrow f(1) = (1)^2 + 1 = 1 + 1 = 2 \neq 4$

C. $f(x) = \sqrt{x} + 3$, if $x = 1 \rightarrow f(1) = \sqrt{1} + 3 = 4$

D. $f(x) = 3x^2 + 1$, if $x = 1 \rightarrow f(1) = 3(1)^2 + 1 = 4$

Answers A and B are not correct and one of the answers C and D is correct. Review the choices when $x = 2$.

C. $f(x) = \sqrt{x} + 3$, if $x = 2 \rightarrow f(2) = \sqrt{2} + 3 \neq 4$

D. $f(x) = 3x^2 + 1$, if $x = 2 \rightarrow f(2) = 3(2)^2 + 1 = 13$

Therefore, answer D is correct.

9) Choice A is correct.

Find the corresponding sides and write a proportion: $\frac{AB}{BC} = \frac{EF}{FG}$

Substitute 13 for AB, 10.3 for BC, and 7 for EF.

Then: $\frac{13}{10.3} = \frac{7}{FG}$

10) Choice D is correct.

Substitute numbers $1, -2, 0, 3, -4, 10$ for x:

$x = 1 \rightarrow 4(1) - 3 > 2 \rightarrow 1 > 2$, This is not true.

$x = -2 \rightarrow 4(-2) - 3 > 2 \rightarrow -11 > 2$, This is not true.

$x = 0 \rightarrow 4(0) - 3 > 2 \rightarrow -3 > 2$, This is not true.

$x = 3 \rightarrow 4(3) - 3 > 2 \rightarrow 9 > 2$, This is true.

$x = -4 \rightarrow 4(-4) - 3 > 2 \rightarrow -19 > 2$, This is not true.

$x = 10 \rightarrow 4(10) - 3 > 2 \rightarrow 37 > 2$, This is true.

So, choice D is correct.

11) The answer is 90.

The surface area of the figure is equal to the sum of the areas of two triangles and three rectangles.

This equals $3 \times 14 \times 2 + 2 \times \frac{1}{2} \times 3 \times 2 = 84 + 6 = 90\ cm^2$

12) Choice B is correct.

At Store P, the shoes will be discounted by 30%, which is $45 (30% of $150). Therefore, the sale price would be $150 − $45 = $105.

At Store Q, the shoes will be discounted by a flat rate of $50. Therefore, the sale price would be $150 − $50 = $100.

Both stores offer discounts, but Store P's discount is based on a percentage off the original price, which means that the discount amount will increase or decrease depending on the price of the shoes. Store Q's discount, on the other hand, is a fixed amount, which may not be as beneficial if the original price of the shoes is higher.

13) Choice B is correct.

We substitute $1, 2, 3, 4, , ...$ for x in the expression $4x + 6$:

$x = 1 \rightarrow 4(1) + 6 = 10$

$x = 2 \rightarrow 4(2) + 6 = 14$

$x = 3 \rightarrow 4(3) + 6 = 18$

$x = 4 \rightarrow 4(4) + 6 = 22$

So, choice B is correct.

14) Choice A is correct.

From the choices provided only point (4,4) is on the path.

15) The answer is 400.

To solve this problem, we can use the formula:

$$Weight\ of\ Solution = \frac{Weight\ of\ Salt}{Salt\ Concentration}$$

We are given that the solution contains 15% salt and there are 60 g of salt. So, we can substitute these values into the formula and solve for the weight of the solution:

Weight of Solution $= \frac{60\ g}{0.15}$

Weight of Solution $= 400\ g$

Therefore, the weight of the solution is 400 g. So, the correct answer is option C.

16) Choice A is correct.

To find the area of the rectangular room, we need to multiply its length by its width:

$Area = length \times width$

Converting the measurements to meters, we get:

$Length = 10 \, feet \times 0.3048 \, \dfrac{meters}{foot} = 3.048 \, meters$

$Width = 12 \, feet \times 0.3048 \, \dfrac{meters}{foot} = 3.658 \, meters$

$Area = 3.048 \, meters \times 3.658 \, meters = 11.149584 \, square \, meters$

The answer is A $11.15 \, m^2$.

17) Choice C is correct.

To find the number of students who prefer Math to History, we need to calculate the number of students who prefer Math and the number of students who prefer History and then find the difference between them.

From the chart, we can see that 40% of students prefer Math, and 14% of students prefer History.

So, the number of students who prefer Math is: $40\% \, of \, 250 = \left(\dfrac{40}{100}\right) \times 250 = 100$

And the number of students who prefer History is: $14\% \, of \, 250 = \left(\dfrac{14}{100}\right) \times 250 = 35$

The difference between the number of students who prefer Math and History is: $100 - 35 = 65$

Therefore, the answer is option C, 65. There are 65 more students who prefer Math than History.

18) Choice B is correct.

To find the mean of the data, we need to sum all the values and divide by the total number of students:

$Mean = \dfrac{2+3+4+4+5+6+7}{7} = \dfrac{31}{7} \approx 4.43$

To find the mode of the data, we need to identify the value that appears most frequently. In this case, both 4 appear twice.

The difference between the mean and mode is the absolute value of the difference between the mean and the mode. We have: $|Mean - Mode| = |4.43 - 4| = 0.43$

Therefore, the answer is B.

19) The answer is $\frac{64}{343}$.

The square of a number is $\frac{16}{49}$, then the number is the square root of $\frac{16}{49}$: $\sqrt{\frac{16}{49}} = \frac{4}{7}$

And the cube of the number is: $(\frac{4}{7})^3 = \frac{64}{343}$

20) Choice C is correct.

Substitute the value of x and y. $5(2x + 4y) + (3 + x)^2$ when $x = 3$ and $y = 2$.

$5(2x + 4y) + (3 + x)^2 = 5\big(2(3) + 4(2)\big) + (3 + 3)^2 = 5(6 + 8) + (6)^2 = 70 + 36 = 106$

21) Choice B is correct.

We can start by converting the mixed number $4\frac{1}{3}$ to an improper fraction:

$4\frac{1}{3} = \frac{3 \times 4 + 1}{3} = \frac{13}{3}$

Then we can divide -16.9 by $\frac{13}{3}$:

$-16.9 \div \left(\frac{13}{3}\right) = -16.9 \times \frac{3}{13} = -3.9$

Therefore, the answer is B.

22) Choice A is correct.

A reflection is a flip over a line. Notice that each point of the original figure and its image are the same distance away from the line of reflection. Therefore, ABC was reflected across the $x-$axis.

23) Choice A is correct.

The answer is A.

Since the company has a budget of $5,000, the total cost of the computers and printers combined cannot exceed $5,000. The cost of c computers is $1000\,c$ and the cost of p printers is $500\,p$.

Therefore, the inequality that represents all possible values of c and p is:

$1000\,c + 500\,p \le 5000$

This inequality ensures that the company does not exceed its budget while purchasing computers and printers.

24) Choice A is correct.

To calculate the percentage of Mariah's income used to pay for her rent and groceries, we need to add up the amounts for these two categories and divide the total by her monthly income, then multiply by 100 to get the percentage.

Total spent on rent and groceries = $1,800 + $450 = $2,250

Percentage of income used for rent and groceries = $\left(\frac{\$2,250}{\$6,000}\right) \times 100\% = 37.5\%$

Therefore, the answer is A, 37.5% (rounded to the nearest whole number).

25) Choice A is correct.

The equation that can be used to find y, the total amount of money Sophie will have saved after x months is:

A. $y = 100x + 500$

Sophie starts with $500, which is the $y-$intercept of the equation. She saves $100 each month, which is the slope of the equation. Therefore, to find out how much money Sophie will have saved after x months, we simply multiply the number of months (x) by the amount saved per month ($100) and add the initial amount ($500). This gives us the equation:

$y = 100x + 500$

Therefore, option A is the correct answer.

26) Choice C is correct.

To find the total amount Sophie paid, we need to multiply the cost of each box by the number of boxes and add the cost of wrapping.

Cost of chocolates = $12 \text{ boxes} \times \$5.50 \text{ per box} = \$66.00$

$Total\ cost = Cost\ of\ chocolates + Cost\ of\ wrapping = \$66.00 + \$10.25 = \$76.25$

Therefore, Sophie paid a total of $76.25 for 12 boxes of chocolates.

27) Choice C is correct.

Given the fact that the three interior angles of a triangle always add up to $180°$.

Therefore: $(2x) + (3x + 4) + 86 = 180 \rightarrow 5x + 90 = 180 \rightarrow 5x = 90 \rightarrow x = 18$

28) Choice B is correct.

The general slope – intercept form of the equation of a line is $y = mx + b$, where m is the slope and b is the y −intercept.

We select two coordinate points: $A(-3, 1)$ and $B(3, -3)$

Then find the slope: $= \frac{y_2 - y_1}{x_2 - x_1} = \frac{-3 - 1}{3 - (-3)} = \frac{-4}{6} = -\frac{2}{3} \rightarrow m = -\frac{2}{3}$

b is the y −intercept, so $b = -1$.

The equation of the line is: $y = -\frac{2}{3}x - 1$

29) Choice C is correct.

Let's assume the number of blue marbles in the box to be $3x$, and the number of green marbles to be $4x$. Therefore, the total number of marbles in the box is $7x$.

We are given that the total number of blue and green marbles is 210, so we can set up the following equation: $3x + 4x = 210$

$3x + 4x = 210$ Simplifying this equation, we get: $7x = 210 \rightarrow x = 30$

So, the number of blue marbles in the box is: $3x = 3(30) = 90$

$3x = 3(30) = 90$ Therefore, there are 90 blue marbles in the box.

Hence, the answer is option C, 90.

30) Choice B is correct.

To calculate the sales tax on the restaurant bill, we first need to find the tax rate:

$Tax\ rate = 7.5\%\ of\ \$60 = \left(\frac{7.5}{100}\right) \times \$60 = \$4.50$

Therefore, the sales tax on the restaurant bill is $4.50, which is option B.

31) Choice C is correct.

The architect used a scale of 1 inch representing 1 foot, which means that every inch on the blueprint represents 1 foot in reality.

So, if the actual length of the living room is 20 feet, then the length of the living room on the blueprint would be 20 inches.

Therefore, the answer is C, 20 inches.

32) Choice B is correct.

To find the area of the shaded area, you need to subtract the area of the two circles from the area of the rectangle. (S_1: the area of the rectangle. S_2: the area of the two circles). Use the area of circle formula: $S = \pi r^2$

$$r = \frac{6}{2} = 3\,m$$

$$S_1 - S_2 = 12 \times 6 - (2 \times \pi(3)^2) \rightarrow S_1 - S_2 = 72 - 18\pi \rightarrow S_1 - S_2 = 72 - 18\pi\ m^2$$

33) Choice D is correct.

Let x equal the smallest angle of the triangle. Then, the three angles are $2x$, $3x$, and $7x$. The sum of the angles of a triangle is 180. Set up an equation using this to find x:

$$2x + 3x + 7x = 180 \rightarrow 12x = 180 \rightarrow x = 15$$

Since the question asks for the measure of the smallest angle, $2x = 2(15) = 30°$

34) Choice C is correct.

To find the ratio of the average speeds of Tom and Sarah, we first need to calculate their individual average speeds.

Tom's average speed: $\frac{112\ miles}{8\ hours} = 14\ miles\ per\ hour$

Sarah's average speed: $\frac{80\ miles}{5\ hours} = 16\ miles\ per\ hour$

Now we find the ratio of Tom's average speed to Sarah's average speed: $14 : 16$

We can simplify this ratio by dividing both numbers by their greatest common divisor, which is 2: $\left(\frac{14}{2}\right) : \left(\frac{16}{2}\right) = 7 : 8$

So, the ratio of Tom's average speed to Sarah's average speed is $7 : 8$.

35) Choice A is correct.

To solve the inequality $7x - 12 \leq -8$, we need to isolate x on one side of the inequality sign.

Adding 12 to both sides gives us: $7x \leq 4$

Dividing both sides by 7 gives us: $x \leq \frac{4}{7}$

Therefore, the solution set for the inequality $7x - 12 \leq -8$ is:

A. $x \leq \frac{4}{7}$

36) Choice A is correct.

There are five arrows and each one is equivalent to 2, then: 5. (2)

37) Choice B is correct.

Plug in the values of x and y provided in the choices into both equations. Let's start with $3x + y = 12$:

A. $(7,7)$ $3x + y = 12 \rightarrow 21 + 7 \neq 12$

B. $(7,-9)$ $3x + y = 12 \rightarrow 21 + (-9) = 12$

C. $(3,5)$ $3x + y = 12 \rightarrow 9 + 5 \neq 12$

D. $(1,2)$ $3x + y = 12 \rightarrow 3 + 2 \neq 12$

Only choice B is correct.

38) Choice C is correct.

To find the constant of proportionality, we need to divide the total number of servings by the amount of sugar used in the recipe.

$$Constant\ of\ proportionality = \frac{Total\ number\ of\ servings}{Amount\ of\ sugar\ used}$$

$$Constant\ of\ proportionality = \frac{24}{4} = 6$$

Therefore, the constant of proportionality that relates the number of servings, y, to the number of cups of sugar used, x, is 6. The answer is C.

39) Choice A is correct.

If Jenny has downloaded 240 songs which take up 720 megabytes of space, then each song takes up $\frac{720}{240} = 3$ megabytes of space.

Therefore, 30 songs would take up $30 \times 3 = 90$ megabytes of space.

So, the answer is A, 90 MB.

40) Choice D is correct.

To find the total distance Samantha traveled, we need to calculate the distance traveled during each segment and then add them up.

Distance traveled during the first segment of cycling = $Speed \times Time$

$$= 12\frac{km}{h} \times 0.5\,h = 6\,km$$

Distance traveled during the jogging segment = Speed × Time

$$= 6\frac{km}{h} \times 0.75h = 4.5km$$

Distance traveled during the second segment of cycling = Speed × Time

$$= 10\frac{km}{h} \times 0.33h = 3.3km$$

Total distance traveled $= 6km + 4.5km + 3.3km = 13.8km$

Therefore, Samantha traveled a total distance of $13.8\,km$ during this time. The answer is D.

STAAR Mathematics Practice Test 3
Answers and Explanations

1) Choice C is correct.

To find the dimensions of a triangle that is similar to triangle QRS, we need to maintain the same ratios of corresponding sides as in QRS. Looking at the given dimensions of QRS, we can see that the ratio of the sides is $8:15:17$.

Let's try each option to see if they maintain this ratio:

A. $6\ cm, 10.5\ cm, 13.5\ cm$

The ratio of sides would be $6:10.5:13.5$, which simplifies to $4:7:9$. This is not the same ratio as in QRS, so this option is not correct.

B. $7cm, 14\ cm, 17.5\ cm$

The ratio of sides would be $7:14:17.5$, which simplifies to $2:4:5$. This is not the same ratio as in QRS, so this option is not correct.

C. $4\ cm, 7.5\ cm, 8.5\ cm$

The ratio of sides would be $4:7.5:8.5$, which simplifies to $8:15:17$. This is the same ratio as in QRS, so this option is correct.

D. $9\ cm, 18\ cm, 27cm$

The ratio of sides would be $9:18:27$, which simplifies to $1:2:3$. This is not the same ratio as in QRS, so this option is not correct.

Therefore, the answer is C, $4\ cm, 7.5\ cm, 8.5\ cm$.

2) Choice A is correct.

To determine the brand of cereal that has the highest amount of sugar per ounce, we need to calculate the amount of sugar in grams per ounce for each brand.

For Brand A, the amount of sugar per ounce is $12\frac{g}{6}oz = 2\frac{g}{oz}$

For Brand B, the amount of sugar per ounce is $16\frac{g}{16}oz = 1g/oz$

For Brand C, the amount of sugar per ounce is $8\frac{g}{10}oz = 0.8g/oz$

For Brand D, the amount of sugar per ounce is $10\frac{g}{14}oz = 0.71g/oz$

The answer is A.

3) Choice B is correct.

The equation that can be used to find y, the total cost of a ride that is x miles long, is:

B. $y = 2x + 20$

In this equation, $2x$ represents the additional cost based on the number of miles driven, and 20 represents the flat fee. The equation can be read as "*y is equal to* 2 *times x plus* 20".

4) Choice B is correct.

To calculate the percentage of Samantha's income that goes towards housing and food, we need to add up the amounts for these two categories and then divide by her total income, and multiply by 100 to get the percentage.

Total amount for housing and food = \$1,500 + \$750 = \$2,250

Percentage of income spent on housing and food $= \left(\frac{\$2,250}{\$4,500} \right) \times 100\% = 50\%$

Therefore, the answer is B, 50%. Samantha spends around 50% of her monthly income on housing and food.

5) Choice D is correct.

The number of wristbands sold can be calculated by subtracting the number of T-shirts sold from the total number of items sold:

Number of wristbands sold

$= Total\ number\ of\ items\ sold - Number\ of\ T-shirts\ sold$

Number of wristbands sold $= 200 - 120 = 80$

The money raised from selling wristbands can be calculated by multiplying the number of wristbands sold by the price per wristband:

Money raised from selling wristbands

$= Number\ of\ wristbands\ sold \times Price\ per\ wristband$

Money raised from selling wristbands $= 80 \times \$2 = \$160$

Therefore, the sports team raised \$160 from selling wristbands. The answer is D, \$160.

6) Choice D is correct.

The total weight of the units produced in that week can be calculated by multiplying the number of units by the weight of each unit:

$4,500 \ units \times 2.5 \ \frac{lbs}{unit} = 11,250 \ lbs$

So, the closest option to the total weight of the units produced in that week is D, 11,250 lbs.

7) Choice C is correct.

To find the probability of getting at least one head and one tail in a three-coin flip, we can subtract the probability of getting all heads or all tails from 1.

The probability of getting all tails is $\frac{7}{50}$, and the probability of getting all heads is $\frac{11}{50}$.

So, the probability of getting at least one head and one tail is $1 - \left(\frac{7}{50}\right) - \left(\frac{11}{50}\right) = \frac{32}{50} = \frac{16}{25}$.

To find how many of the next 150 trials will have this outcome, we can multiply the probability by the total number of trials: $\left(\frac{16}{25}\right) \times 150 = 96$

Therefore, the correct answer is 96.

8) Choice B is correct.

To find the sales tax on the bicycle, we need to multiply the cost of the bicycle by the sales tax rate.

$$Sales \ tax = Cost \ of \ bicycle \times Sales \ tax \ rate \rightarrow Sales \ tax = \$350 \times 0.08$$

$$\rightarrow Sales \ tax = \$28.00$$

Therefore, the sales tax on the bicycle is $28.00, which is option B.

9) Choice C is correct.

A $\frac{3}{5}$ probability of precipitation for Saturday corresponds to a 60% chance of rain. This percentage indicates that there is a greater chance of rain than no rain, and therefore it can be considered quite likely that it will rain on that day.

In this scenario, option C is the most fitting description, as it states that it is quite likely that it will rain on Saturday. A 60% chance of rain is not a certainty, but it is a significant likelihood, making it more appropriate to describe the situation as quite likely rather than minimal (option B), extremely unlikely (option A), or absolutely certain (option D).

10) The answer is 314.

The formula for the circumference of a circle is $C = 2\pi r$, where π is approximately 3.14 and r is the radius. Solving for the radius, we get $r = \frac{C}{2\pi} = \frac{62.8}{2\pi} \approx 10$. We can then use the formula for the area of a circle, $A = \pi r^2$, to find the area. Substituting the value for r, we get $A = 3.14 \times 10^2 = 314$ square inches.

11) Choice D is correct.

We can use proportions to estimate the number of bottles with a defective cap. The proportion of bottles with a defective cap in the sample is $\frac{12}{200} = 0.06$. We can use this proportion to estimate the number of bottles with a defective cap in the population as follows:

Number of bottles with a defective cap $= 0.06 \times 5,000 = 300$

Thus, we can predict that approximately 300 bottles out of the 5,000 will have a defective cap.

12) Choice C is correct.

Let's start by finding out how long it took Sarah to complete 60% of her walk.

If Sarah completed 60% of her walk in $2\frac{1}{4}$ hours, that means she has 40% left to complete.

To find out how long it will take her to complete the remaining 40%, we can use the following proportion:

$$\frac{60\%}{2\frac{1}{4}hours} = \frac{40\%}{x}$$

where x is the amount of time it will take Sarah to complete the remaining 40% of her walk.

To solve for x, we can cross-multiply and simplify:

$60\% \times x = 40\% \times \frac{9}{4} \ hours \rightarrow 0.6x = 0.4 \times \frac{9}{4} \rightarrow 0.6x = 0.9 \rightarrow x = \frac{0.9}{0.6} \rightarrow x = \frac{3}{2} \ hours$

So, it will take Sarah an additional $\frac{3}{2}$ hours to complete the remaining 40% of her walk.

To find the total time it will take her to complete her entire walk, we can add the time it took her to complete the first 60% ($\frac{9}{4}$ hours) to the time it will take her to complete the remaining 40% ($\frac{3}{2}$hours):

$$\frac{9}{4} + \frac{3}{2} = \frac{15}{4} = 3\frac{3}{4}$$

Sarah's entire walk will take $3\frac{3}{4}$ hours, which is answer choice C.

13) Choice A is correct.

We can use the data in the table to find the relationship between the number of pages read and the time taken. We can see that Maria's reading rate is consistent, which means that she reads at a constant speed. Therefore, we can use proportions to find the equation that relates pages read and time taken.

Let p be the number of pages read and t be the time taken in minutes. Then, for each row in the table, we have: $\frac{60}{45} = \frac{p}{t}, \frac{80}{60} = \frac{p}{t}, \frac{100}{75} = \frac{p}{t}, \frac{120}{90} = \frac{p}{t}$

Simplifying each equation, we get: $\frac{4}{3} = \frac{p}{t}, \frac{4}{3} = \frac{p}{t}, \frac{4}{3} = \frac{p}{t}, \frac{4}{3} = \frac{p}{t}$

Therefore, we can conclude that the correct equation is A, $p = \left(\frac{4}{3}\right)t$.

14) Choice B is correct.

Let's first find the length of side QR:

Side $QR = 1.2 \times PQ = 1.2 \times 16 = 19.2$ inches

Then, we can find the length of side PR:

Side $PR = \left(\frac{3}{4}\right) \times QR = \left(\frac{3}{4}\right) \times 19.2 = 14.4$ inches

Now, we need to convert the lengths to feet, using the scale factor of $\frac{1}{4}$:

$$PQ = \frac{16}{4} = 4 \, feet$$

$$QR = \frac{19.2}{4} = 4.8 \, feet$$

$$PR = \frac{14.4}{4} = 3.6 \, feet$$

Finally, we can calculate the perimeter of the actual triangular face:

$$Perimeter = PQ + QR + PR = 4 + 4.8 + 3.6 = 12.4 \, feet$$

Therefore, the correct answer is B, 12.4.

15) Choice D is correct.

When a square pyramid is cut through the vertex and perpendicular to the base, the cross section created is a triangle. This triangle is an isosceles triangle since it is formed by the intersection of two congruent triangles. Therefore, the correct answer is D, an isosceles triangle.

16) The answer is 54°.

According to the graph, 15% of people are interested in jazz music. Find 15% of 360°.

$15\% \ of \ 360° = 0.15 \times 360° = 54°$

17) Choice A is correct.

The general slope-intercept form of the equation of a line is $y = mx + b$, where m is the slope and b is the $y -$intercept. Select two coordinate points: $A(-1, -5)$ and $B(3, 3)$

Then find the slope: $= \frac{y_2 - y_1}{x_2 - x_1} = \frac{-5 - 3}{-1 - 3} = \frac{-8}{-4} = 2 \rightarrow m = 2$

b is the $y -$intercept, so $b = -3$. The equation of the line is: $y = 2x - 3$

18) Choice D is correct.

To arrive at the most reliable conclusion, the company should use a sample that is representative of the entire population. Option D, 50 employees from each facility, would be the best sample size because it includes employees from each facility and ensures that the sample is large enough to provide reliable data.

19) Choice C is correct.

The radius of the tank is 5 feet (half of the diameter). The formula for the volume of a cylinder is $V = \pi r^2 h$, where π is approximately 3.14, r is the radius, and h is the height. Substituting the given values, we get $V = 3.14 \times 5^2 \times 20 = 1,570$ cubic feet. To find the amount of water needed to fill the tank to 90% capacity, we multiply the volume of the tank by 0.9, giving us 1,413 cubic feet. Finally, we multiply this result by 7.48 to convert cubic feet to gallons, giving us 10,569.24 gallons.

20) Choice A is correct.

A. The probability Ms. Johnson selects a sophomore on Friday is the same as it was on each of the other days.

Since each student is equally likely to be selected and there is an equal number of freshmen, sophomores, juniors, and seniors, the probability of selecting a sophomore on any given day is $\frac{1}{4}$. The events of selecting a sophomore on each day are independent, so the fact that a sophomore was selected on the previous four days has no impact on the probability of selecting a sophomore on Friday.

21) Choice D is correct.

Area of rectangle, $14 \times 5 = 70 \ m^2$

Area of triangle, $\frac{1}{2}(6) \times (5) = 15 \ m^2$

Area of the semicircular, $\frac{1}{2} \times (\frac{5}{2})^2 \times \pi = 9.81 \ m^2$

The total area of the figure is equal to the sum of the above areas: $70 + 15 + 9.81 = 94.81 \ m^2$

22) Choice C is correct.

The data range of series B is equal to 13 and the data range of series A is equal to 6. So, answer A is not correct.

According to the chart, the interquartile range of group A data is larger than group B data. Therefore, answer B is not correct.

The median of the data of series A and series B is equal to 1, so answer C is correct.

23) The answer is 49.

To solve this problem, we can use the fact that the ratio of cats to dogs is 5 to 7. This means that for every 5 cats, there are 7 dogs.

If there are 35 cats in the shelter, we can find the number of dogs as follows:

Number of dogs = Number of cats × Ratio of dogs to cats

Number of dogs $= \left(\frac{7}{5}\right) \times 35$

Number of dogs $= 49$

Therefore, there are 49 dogs in Mrs. Smith's animal shelter. The answer is C, 49.

24) Choice A is correct.

The correct statement that best describes the height of the roller coaster, given the equation $y = -2x + 10$, is:

A. From a starting height of 10 meters above the ground, the roller coaster is descending 2 meters per meter traveled.

This means that for every meter the roller coaster travels horizontally, it is descending 2 meters vertically, and it starts from a height of 10 meters above the ground.

25) Choice B is correct.

To find the number of green marbles in Katie's collection, we need to subtract the number of red and blue marbles from the total number of marbles.

30% of 900 marbles are red, which is $\left(\frac{30}{100}\right) \times 900 = 270$ marbles.

50% of 900 marbles are blue, which is $\left(\frac{50}{100}\right) \times 900 = 450$ marbles.

Therefore, the total number of red and blue marbles is $270 + 450 = 720$.

To find the number of green marbles, we can subtract the total number of red and blue marbles from the total number of marbles:

Total number of marbles $= 900$

Total number of red and blue marbles $= 720$

Therefore, the number of green marbles in Katie's collection is $900 - 720 = 180$.

So, the answer is B, 180.

26) Choice C is correct.

To find out which equation is true when $x = 5$, we simply substitute 5 for x in each equation and see which equation is true.

A. $5x + 1 = 19$

Substituting $x = 5$, we get:

$5(5) + 1 = 19 \rightarrow 25 + 1 = 19 \rightarrow 26 \neq 19$

This equation is false when $x = 5$.

B. $x - 2 = 7$

Substituting $x = 5$, we get:

$5 - 2 = 7 \rightarrow 3 \neq 7$

This equation is false when $x = 5$.

C. $3x + 7 = 22$

Substituting $x = 5$, we get:

$3(5) + 7 = 15 + 7 \rightarrow 22 = 22$

This equation is true when $x = 5$.

D. $4x - 4 = 14$

Substituting $x = 2$, we get:

$4(5) - 4 = 20 - 4 \rightarrow 16 \neq 14$

This equation is false when $x = 5$.

Therefore, the equation that is true when $x = 5$ is C, $3x + 7 = 22$.

27) Choice A is correct.

To find the total distance Helen covered, we need to calculate the distance traveled during each segment and add them up.

$Distance\ covered\ during\ biking = speed \times time = 10\frac{km}{h} \times 0.5\ h = 5\ km$

$Distance\ covered\ during\ jogging = speed \times time = 5\frac{km}{h} \times 1\ h = 5\ km$

$Distance\ covered\ during\ swimming = speed \times time = 2\frac{km}{h} \times 0.75\ h = 1.5\ km$

$Total\ distance\ covered = 5\ km + 5\ km + 1.5\ km = 11.5\ km$

Therefore, the total distance Helen covered during this time is $11.5\ km$.

28) Choice A is correct.

Maggie needs 2 cups of flour for every 12 cupcakes, so for 120 cupcakes she will need:

$\left(\frac{120}{12}\right) \times 2 = 20$ cups of flour

Maggie has 10 cups of flour in her pantry, so she needs to buy:

$20 - 10 = 10$ cups of flour.

Therefore, the answer is A, Maggie needs to buy 10 more cups of flour.

29) Choice B is correct.

To solve this problem, we first need to find out how many T-shirts and hats Ms. Johnson sells per day.

Number of t-shirts sold per day = 100 t-shirts ÷ 2 days = 50 t-shirts per day

Number of hats sold per day = 75 hats ÷ 2 days = 37.5 hats per day

Now we can find out how many t-shirts and hats she will sell in 6 days:

Number of t-shirts sold in 6 days = 50 T-shirts per day × 6 days = 300 T-shirts

Number of hats sold in 6 days = 37.5 hats per day × 6 days = 225 hats

Therefore, Ms. Johnson will sell $300 - 225 = 75$ more T-shirts than hats in 6 days.

The answer is B, 75.

30) Choice B is correct.

The two statistical measures that are most appropriate for describing the central tendency and variability of the given data set are:

B. Mean and standard deviation.

The mean is the average of the daily sales and gives a measure of the central tendency of the data set. The standard deviation is a measure of the variability or spread of the data around the mean. Together, these two measures provide a good summary of the typical sales level and the extent to which sales vary from day to day.

31) Choice D is correct.

We can start by converting the mixed number $4\frac{2}{5}$ to an improper fraction:

$$4\frac{2}{5} = \frac{4 \times 5 + 2}{5} = \frac{22}{5}$$

Then we can divide 32.25 by $\frac{22}{5}$:

$$32.25 \div \left(4\frac{2}{5}\right) = 32.25 \times \frac{5}{22} \approx 7.32$$

Therefore, the answer is D, 7.32.

32) Choice A is correct.

To represent the budget constraint of the school for the field trip, we can use the following inequality:

$20s + 500 \leq 1,500$

This inequality states that the total cost of the field trip, which includes the cost of admission per student and the cost of renting a bus, should be less than or equal to the available budget of $1,500. Therefore, the school can take all possible values of s, the number of students, that satisfy this inequality.

So, the answer is A, $20s + 500 \leq 1,500$.

33) Choice D is correct.

To solve the inequality $3x + 5 \leq 26$, we need to isolate x on one side of the inequality sign.

Subtracting 5 from both sides gives us: $3x \leq 21$

Dividing both sides by 3 gives us:

$x \leq 7$

Therefore, the solution set for the inequality $3x + 5 \leq 26$ is D.

34) Choice A is correct.

The correct sequence that represents an arithmetic progression with a common difference of 2 and a first term of 9 is A.

In sequence A, we can see that each term is obtained by adding 2 to the previous term. The first term is 9, and if we add 2 to it, we get the second term of 11. Similarly, if we add 2 to 11, we get the third term of 13, and so on. Therefore, sequence A is an arithmetic progression with a common difference of 2 and a first term of 9.

35) The answer is 12.

To find the median of a set of numbers, we need to arrange them in numerical order and then identify the middle number(s). If there is an odd number of values, the median is the middle number, and if there is an even number of values, the median is the average of the two middle numbers.

So, first, we arrange the given numbers in numerical order: $1, 3, 9, 11, 13, 19, 26, 31$

There are eight numbers in the set, which is an even number, so we need to find the average of the two middle numbers. The middle two numbers are 11 and 13, so we add them together and divide by 2: $\frac{11+13}{2} = 12$

Therefore, the median of the given numbers is 12.

36) Choice B is correct.

Count and write each set of tiles, $7x + 4 = 2x + 3$. Solve the resulting equation to determine the value of x:

$7x + 4 = 2x + 3 \rightarrow 7x - 2x + 4 = 2x - 2x + 3 \rightarrow 5x + 4 = 3 \rightarrow 5x + 4 - 4 = 3 - 4$

$$\rightarrow 5x = -1 \rightarrow x = -\frac{1}{5}$$

37) Choice A is correct.

We can use the formula for direct proportionality:

$p = k \times q^3$

where k is the constant of proportionality. To find k, we can use the initial values:

$81 = k \times 3^3 \rightarrow k = \dfrac{81}{27} = 3$

Now we can use the value of k to find q when $p = 192$:

$192 = 3 \times q^3 \rightarrow q^3 = 64 \rightarrow q = 4$

Therefore, the answer is A.

38) Choice D is correct.

The original price of the laptop is $800.00, and the new price is $680.00.

The percentage decrease can be calculated as:

$\left(\dfrac{original\ price - new\ price}{original\ price}\right) \times 100\%$

Substituting the given values, we get:

$\left(\dfrac{800.00 - 680.00}{800.00}\right) \times 100\% = 15\%$

Therefore, the price of the laptop was decreased by 15%.

The answer is D, 15%.

39) Choice B is correct.

If $\dfrac{3}{7}$ of a tank contains 168 liters of water, then one tank contains:

1 tank = $\left(\dfrac{7}{3}\right) \times 168 = 392$ liters

Therefore, the total capacity of six such tanks of water is:

6 tanks = $6 \times 392 = 2,352$ liters

40) Choice A is correct.

$-4 \leq x < 1 \rightarrow$ Multiply all sides of the inequality by 5. Then: $-4 \times 5 \leq 5 \times x < 1 \times 5 \rightarrow$
$-20 \leq 5x < 5$. Add 3 to all sides. Then: $\rightarrow -20 + 3 \leq 5x + 3 < 5 + 3 \rightarrow -17 \leq 5x + 3 < 8$
Minimum value of $5x + 3$ is -17.

STAAR Mathematics Practice Test 4
Answers and Explanations

1) Choice B is correct.

A 48% chance of rain on Saturday means that the probability of precipitation is slightly less than half. In this case, it is neither very likely nor very unlikely that it will rain. Therefore, the most appropriate description of the situation is that there is an even chance of rain on Saturday.

Option B best represents the probability of precipitation in this scenario. A 48% chance of rain is not highly unlikely (Option A), likely (Option C), or guaranteed (Option D). Instead, it suggests a relatively balanced probability, leaning slightly towards no rain but still close enough to be considered an even chance.

2)   Choice D is correct.

Using the same formulas as in the previous question, we can solve for the radius: $r = \frac{C}{2\pi} = \frac{37.68}{2\pi} \approx$ 6. We can then use the formula for the area of a circle, $A = \pi r^2$, to find the area. Substituting the value for r, we get $A = 3.14 \times 6^2 = 113.04$ square meters.

3)   Choice B is correct.

Let's first find the length of side QR: Side $QR = 1.3 \times Q = 1.3 \times 20 = 26$ inches.

Then, we can find the length of side PR: Side $PR = \left(\frac{1}{2}\right) \times QR = \left(\frac{1}{2}\right) \times 26 = 13$ inches.

Now, we need to convert the lengths to feet, using the scale factor of $\frac{1}{3}$:

$PQ = \frac{20}{3} = 6.67$ feet

$QR = \frac{26}{3} = 8.67$ feet

$PR = \frac{13}{3} = 4.33$ feet

Finally, we can calculate the perimeter of the actual triangular face:

Perimeter $= PQ + QR + PR = 6.67 + 8.67 + 4.33 = 19.67$ feet Rounding to the nearest tenth, we get 19.7 feet. Therefore, the correct answer is B.

4) Choice A is correct.

To find the equation that represents the linear relationship between the x −values and the y −values in the given table, we can use the slope-intercept form of a linear equation: $y = mx + b$, where m is the slope and b is the y −intercept.

To find the slope, we can use any two points from the table and calculate the difference in y −values over the difference in x −values:

$$slope = \frac{y_2 - y_1}{x_2 - x_1}$$

Using the points $(0, -3)$ and $(3, 12)$, we get:

$$slope = \frac{12 - (-3)}{3 - 0} = \frac{15}{3} = 5$$

Now that we have the slope, we can use any point from the table and the slope to find the y −intercept. Using the point $(0, -3)$, we get:

$$y = mx + b \rightarrow -3 = 5 \times 0 + b \rightarrow b = -3$$

Therefore, the equation that represents the linear relationship between the x −values and the y −values in the given table is: $y = 5x - 3$

5) Choice A is correct.

Mean would be the most useful measure of central tendency to describe the performance of the class on this test.

The mean score can provide an idea of the average performance of the class, which is calculated by adding up all the scores and dividing by the number of scores.

6) Choice C is correct.

To find the cost of goods sold, we need to use the formula:

$Net\ Income = Revenue - COGS - Expenses$

Plugging in the given values:

$\$10,000 = \$50,000 - COGS - (\$5,000 + \$2,500 + \$25,000)$

$\$10,000 = \$50,000 - COGS - \$32,500$

$\$10,000 = \$17,500 - COGS$

$COGS = \$7,500$

Therefore, the cost of goods sold for this small business is $7,500. The answer is C.

7) Choice A is correct.

There are 25 employees who prefer a sports event.

To determine the best prediction of the preferences of all 50 employees, we can use proportional reasoning. If 5 out of 10 employees surveyed prefer a sports event, we can estimate that 25 out of 50 employees in total would prefer a sports event (since 5 is half of 10, we can assume that 25 is also half of 50). Similarly, if 3 out of 10 employees surveyed prefer a board game event, we can estimate that 15 out of 50 employees in total would prefer a board game event (since 3 is $\frac{3}{10}$ of 10, we can assume that 3 is also $\frac{3}{10}$ of 50). Finally, if 2 out of 10 employees surveyed prefer a cooking event, we can estimate that 10 out of 50 employees in total would prefer a cooking event (since 2 is $\frac{1}{5}$ of 10, we can assume that 2 is also $\frac{1}{5}$ of 50).

Therefore, the best prediction is that there are 25 employees who prefer a sports event.

8) Choice C is correct.

To find the number of non-fiction books in Sarah's collection, we need to subtract the number of biographies and fiction books from the total number of books:

Number of biographies $= 20\% \ of \ 800 = 0.2 \times 800 = 160$

Number of fiction books $= 40\% \ of \ 800 = 0.4 \times 800 = 320$

Total number of non-fiction books $= 800 - 160 - 320 = 320$

Therefore, the answer is C, 320.

9) The answer is 90.

To find the area of the shaded area, you need to subtract the area of the triangle from the area of the rectangle:

Triangle area $= \frac{1}{2} \times 18 \times 10 = 90 \ m^2$

Rectangle area $= 18 \times 10 = 180 \ m^2$

The area of the shaded are is: $180 - 90 = 90 \ m^2$

10) Choice B is correct.

The values of d that make the inequality $-5d + 5\frac{1}{2} \le 17$ true are $d \ge -2\frac{3}{10}$, or d is greater than or equal to -2.3. This can be determined by solving the inequality for d. To do this, we first isolate d by subtracting $5\frac{1}{2}$ from both sides of the inequality:

$$-5d + 5\frac{1}{2} - 5\frac{1}{2} \le 17 - 5\frac{1}{2} \rightarrow -5d \le 11\frac{1}{2}$$

Next, divide both sides of the inequality by -5 to get the solution in terms of d:

$$d \ge -2\frac{3}{10}$$

So, the solution set for the inequality is $d \ge -2\frac{3}{10}$, which means that all values of d that are greater than or equal to -2.3 satisfy the inequality.

11) Choice B is correct.

This pyramid consists of four similar triangles and a square. Therefore, the surface area of the pyramid is equal to the sum of the area of the triangles and the square:

Area of each triangle: $\frac{1}{2} \times 14 \times 8 = 56 \ cm^2$

Square area: $8 \times 8 = 64 \ cm^2$

The surface area of the pyramid: $(4 \times 56) + 64 = 288 \ cm^2$

12) Choice A is correct.

The correct answer is option A, $1, 2, 4, 8, 16, \ldots$

To see why, let's plug in some values of x and see what we get:

When $x = 1, 2^{x-1} = 2^{1-1} = 2^0 = 1$

When $x = 2, 2^{x-1} = 2^{2-1} = 2^1 = 2$

When $x = 3, 2^{x-1} = 2^{3-1} = 2^2 = 4$

When $x = 4, 2^{x-1} = 2^{4-1} = 2^3 = 8$

When $x = 5, 2^{x-1} = 2^{5-1} = 2^4 = 16$

As you can see, each term is twice the previous term, which means that the sequence is a geometric sequence with a common ratio of 2. Therefore, the sequence represented by the expression 2^{x-1} is $1, 2, 4, 8, 16, \ldots$ which is option A.

13) Choice D is correct.

To find the probability that Sarah's pizza does not have olives on it, we need to add up the percentages of customers who ordered all the toppings except olives.

Probability of pizza without olives = Probability of pizza with pepperoni + Probability of pizza with mushrooms + Probability of pizza with onions = 45% + 30% + 5% = 80%

Therefore, the probability that Sarah's pizza does not have olives on it is 0.80 or 80%.

14) Choice A is correct.

A. The probability Mr. Patel selects a female customer on Friday is the same as it was on each of the other days.

Explanation: Since each customer is equally likely to be selected and there is an equal number of male and female customers, the probability of selecting a female customer on any given day is $\frac{1}{2}$.

The events of selecting a female customer on each day are independent, so the fact that a female customer was selected on the previous four days has no impact on the probability of selecting a female customer on Friday.

15) Choice C is correct.

We can use the data in the table to find the relationship between the number of pages read and the time taken. We can see that Alex's reading rate is consistent, which means that he reads at a constant speed. Therefore, we can use proportions to find the equation that relates pages read and time taken.

Let p be the number of pages read and t be the time taken in minutes. Then, for each row in the table, we have:

$$\frac{30}{20} = \frac{p}{t} \rightarrow \frac{3}{2} = \frac{p}{t}$$

$$\frac{45}{30} = \frac{p}{t} \rightarrow \frac{3}{2} = \frac{p}{t}$$

$$\frac{60}{40} = \frac{p}{t} \rightarrow \frac{3}{2} = \frac{p}{t}$$

$$\frac{75}{50} = \frac{p}{t} \rightarrow \frac{3}{2} = \frac{p}{t}$$

Therefore, we can conclude that the correct equation is C, $p = \left(\frac{3}{2}\right)t$.

16) Choice B is correct.

We can use proportions to estimate the number of boxes with a broken seal. The proportion of boxes with broken seals in the sample is $\frac{20}{500} = 0.04$. We can use this proportion to estimate the number of boxes with a broken seal in the population as follows:

Number of boxes with a broken seal $= 0.04 \times 10,000 = 400$

Thus, we can predict that approximately 400 boxes out of the 10,000 will have a broken seal.

17) Choice A is correct.

To find the minimum amount of money the student must earn annually to meet this budget, we need to calculate the total monthly budget and then multiply it by 12 to get the annual budget.

Total monthly budget $= \$500 + \$200 + \$250 + \$100 + \$150 + \$50 = \$1,250$

Therefore, the equation to find the minimum amount of money the student must earn annually to meet this budget is: $b = \$1,250 \times 12$

18) Choice D is correct.

Simplifying the fractions, the unit rate of cups per teaspoon is $\frac{3}{5} \div \frac{2}{3} = \frac{9}{10}$ cups per teaspoon.

19) Choice D is correct.

The formula for the volume of a sphere is $V = \left(\frac{4}{3}\right)\pi r^3$, where π is approximately 3.14 and r is the radius. Substituting the given value, we get $V = \left(\frac{4}{3}\right) \times 3.14 \times 2^3 = 33.49$ cubic feet. To find the amount of helium needed to fill the balloon to 75% capacity, we multiply the volume of the balloon by 0.75, giving us 25.12 cubic feet. Finally, we multiply this result by 0.0118 to convert cubic feet to pounds, giving us 0.296 pounds of helium to the nearest hundredth.

20) Choice A is correct.

The cost of textbooks is \$50 each and the cost of workbooks is \$10 each. Therefore, the total cost of textbooks and workbooks can be represented as $50t + 10w$.

Since the school has a budget of \$10,000, the inequality that represents all possible values of t and w is: $50t + 10w \leq 10,000$

This inequality ensures that the total cost of textbooks and workbooks purchased by the school does not exceed the budget of \$10,000. Therefore, option A is the correct answer.

21) Choice D is correct.

To arrive at the most reliable conclusion, the company should use a sample that is representative of the entire population. Option D, 50 employees from each store, would be the best sample size because it includes employees from each store and ensures that the sample is large enough to provide reliable data.

22) Choice C is correct.

Each marble color has an expected frequency of $\frac{8}{4} = 2$. The observed frequency for red is 1, for blue it is 2, for green it is 3, and for yellow it is 2. Thus, the marble color whose observed frequency is closest to its expected frequency is C, Blue.

23) Choice C is correct.

Find the points one by one on the coordinate plane to see which one is inside the circle. For example, the point $(3, -2)$ is 3 units to the right of the x −axis and 2 units below the y −axis. Find the rest of the points in the same way on the coordinate plane. After finding the points, you can see that only answer C is inside the circle.

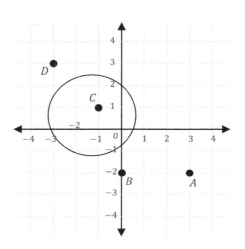

24) Choice B is correct.

The scale of the map is 1 inch represents 2 miles. So, for every 2 miles in the real world, there is 1 inch on the map.

If the actual distance between two locations is 8 miles, then the distance between them on the map can be found by multiplying 8 by the scale of the map, which is 1 inch per 2 miles:

$$8 \ miles \times \left(\frac{1 \ inch}{2 \ miles} \right) = 4 \ inches$$

Therefore, the distance between the two locations on the map is 4 inches.

So, the answer is B, 4 inches.

25) Choice D is correct.

Hannah has a total of $8 + 3 + 5 = 16$ fruits in her basket. The probability of her choosing an apple is given by:

$$\frac{Number\ of\ apples}{Total\ number\ of\ fruits\ in\ the\ basket}$$

So, the probability of Hannah choosing an apple is: $\frac{3}{16}$

Therefore, the probability of Hannah choosing an apple from her basket is $\frac{3}{16}$.

26) The answer is 30.

To solve this problem, we need to find the number of chocolate cupcakes, given the ratio of chocolate to vanilla cupcakes and the total number of cupcakes.

Let C be the number of chocolate cupcakes and V be the number of vanilla cupcakes.

According to the problem, the ratio of chocolate to vanilla cupcakes is 3 to 5. So, we can write:
$\frac{C}{V} = \frac{3}{5}$

We can simplify this by cross-multiplying: $5C = 3V$

We also know that the total number of cupcakes is 80. So: $C + V = 80$

We can use the equation $5C = 3V$ to eliminate V and express C in terms of the total number of cupcakes:

$5C = 3V \rightarrow 5C = 3(80 - C) \rightarrow 5C = 240 - 3C \rightarrow 8C = 240 \rightarrow C = 30$

Therefore, there are 30 chocolate cupcakes in the bakery.

The answer is B, 30.

27) Choice A is correct.

The equation of a line in slope intercept form is: $y = mx + b$. Solve for y:

$y - 5x = 4 \Rightarrow y = 5x + 4$

The slope of this line is 5. Parallel lines have the same slopes.

28) Choice D is correct.

The area of a triangle: $\frac{1}{2} \times base \times height = \frac{1}{2} \times 4\,m \times 3\,m$

We know that $1\,m = 100\,cm \rightarrow \frac{1}{2} \times 400\,cm \times 300\,cm = 60,000\,cm^2 = 100\,(600)\,cm^2$

Only choice D equals to $60,000\,cm^2$

$60,000 \ cm^2 = 100 \ (600) cm^2$

29) The answer is 25.

To solve this problem, we can use the scale factor between the scale drawing and the actual pool dimensions, which is:

$$scale \ factor = \frac{actual \ length}{length \ in \ scale \ drawing} = \frac{40 \ ft}{8 \ in} = 5 \frac{ft}{in}$$

We can use this scale factor to find the actual width of the pool.

$actual \ width = width \ in \ scale \ drawing \times scale \ factor$

$actual \ width = 5 \ in \times 5 \frac{ft}{in}$

$actual \ width = 25 \ ft$

Therefore, the actual width of the pool is 25 feet, which is option C.

30) Choice A is correct.

To find the answer, we can use the formula:

$time = distance \div speed$

Plugging in the given values, we get:

$time = 490 \ miles \div 70 \ miles \ per \ hour$

$time = 7 \ hours$

Therefore, it would take 7 hours to arrive at the destination while driving at an average speed of 70 miles per hour on a journey that is 490 miles long. The answer is option A.

31) Choice B is correct.

To calculate the interest earned on Bob's savings account, we can use the formula:

$Interest = Principal \times Rate \times Time$

Where:

Principal = $10,000 (Bob's balance)

Rate = 2.5% or 0.025 (in decimal form)

Time = 1 year

So, the interest earned by Bob at the end of one year would be:

Interest = $10,000 × 0.025 × 1 = $250

Therefore, the answer is option B, $250.00.

32) Choice A is correct.

To find the percentage of Lila's income used for transportation and entertainment, we need to add up the amounts for those two categories, which is $350 + $300 = $650.

Then, we divide that total by Lila's monthly income and multiply by 100 to get the percentage:

($650 ÷ $5,000) × 100 = 0.13 × 100 = 13%

Therefore, the answer is option A, 13%.

33) Choice D is correct.

The area of the middle trapezoid, $\frac{1}{2} \times (6 + 8) \times (12) = 84\ m^2$

Area of right trapezoid, $\frac{1}{2} \times (8 + 14) \times (5) = 55\ m^2$

Area of the semicircle, $\frac{1}{2} \times \pi \times (3)^2 = 14.13\ m^2$

The total area of the figure is equal to the sum of the above areas:

$84 + 55 + 14.13 = 153.13\ m^2$

34) Choice C is correct.

To find the area of the pool in square meters, we first need to convert the dimensions from feet to meters.

$25\ feet \times 0.3048\ \dfrac{meters}{foot} = 7.62\ meters$

$50\ feet \times 0.3048\ \dfrac{meters}{foot} = 15.24\ meters$

The area of the pool is: $7.62\ meters \times 15.24\ meters = 116.12\ square\ meters$

Therefore, the answer is option C, $116.12\ m^2$.

35) Choice C is correct.

Since triangle ABC is an isosceles triangle, the measure of angle B is also equal to the measure of angle C. Therefore, we can write: $angle\ B = angle\ C = 3x + 10$

The sum of the angles of a triangle is always 180°. Therefore, we can write:

$angle\ A + angle\ B + angle\ C = 180$

Substituting the given values, we get: $70 + (3x + 10) + (3x + 10) = 180$

Simplifying this equation, we get: $6x + 90 = 180$

36) Choice A is correct.

Solve for x. $-3 < 4x - 3 \leq 1 \Rightarrow$ (add 3 all sides) $-3 + 3 \leq 4x - 3 + 3 < 1 + 3 \rightarrow$
$0 \leq 4x < 4 \rightarrow$ (Divide all sides by 4) $0 \leq x < 1$.

x is between 0 and 1. Choice A represents this inequality.

37) Choice A is correct.

The correct equation to find the total amount of money Maggie will have saved after x months is
A. $y = 200x + 1,000$.

Maggie starts with \$1,000, and saves \$200 each month, so after x months, the total amount she will have saved is $200x + 1,000$. Similarly, Maggie starts with \$1,000 and saves \$200 each month, so after x months, the total amount she will have saved is $200x + 1,000$. Therefore, the correct equation is $y = 200x + 1,000$.

38) The answer is $\frac{1}{4}$.

There are 13 hearts in a standard 52-card deck, so the probability of selecting a heart from the deck is $\frac{13}{52}$ or $\frac{1}{4}$.

Therefore, the answer is option B.

39) Choice B is correct.

To manufacture 500 toys, the company uses 3 bags of stuffing material.
So, to manufacture 750 toys, the company would require bags of stuffing material in the same proportion.

bags of stuffing material $= \left(\frac{750}{500}\right) \times 3$

bags of stuffing material $= 4.5$

Therefore, the company needs 4.5 bags of stuffing material to manufacture 750 toys. However, since it's not possible to have half a bag of stuffing material, the company would need to round up to the nearest whole number, so the answer is 5.

40) Choice C is correct.

To find the number of units the company will produce in 5 days, we first need to find the average daily production rate.

Total production in 3 days $= 150 + 175 + 225 = 550$ units

Average daily production rate $= \frac{Total\ production}{Number\ of\ days} = \frac{550}{3} = 183.33$ units per day (rounded to 2 decimal places)

Therefore, the company will produce $183.33 \times 5 = 916.67$ units in 5 days.

So, the answer is option C, 916.67 units.

STAAR Mathematics Practice Test 5
Answers and Explanations

1) Choice A is correct.

We can use the data in the table to find the relationship between the number of pages read and the time taken. We can see that David's reading rate is consistent, which means that he reads at a constant speed. Therefore, we can use proportions to find the equation that relates pages read and time taken.

Let p be the number of pages read and t be the time taken in minutes. Then, for each row in the table, we have:

$$\frac{40}{30} = \frac{p}{t} \rightarrow \frac{4}{3} = \frac{p}{t}$$

$$\frac{60}{45} = \frac{p}{t} \rightarrow \frac{4}{3} = \frac{p}{t}$$

$$\frac{80}{60} = \frac{p}{t} \rightarrow \frac{4}{3} = \frac{p}{t}$$

$$\frac{100}{75} = \frac{p}{t} \rightarrow \frac{4}{3} = \frac{p}{t}$$

Therefore, we can conclude that the correct equation is A, $p = \left(\frac{4}{3}\right)t$.

2) Choice C is correct.

First, find the slope of the line using the formula. To use the formula, you must consider two points on the line, for example $(0, 10)$ and $(11, 0)$.

$$m = \frac{y_2 - y_1}{x_2 - x_1} = \frac{10 - 0}{0 - 11} = -\frac{10}{11}$$

The next step is to find the $y-$intercept of the line. The intersection of the line with the $y-$axis is $(0, 10)$. Therefore, the $y-$intercept is 10. The equation of the line will be as follows:

$$y = mx + b \rightarrow y = -\frac{10}{11}x + 10$$

3) Choice B is correct.

We can solve this problem by using the formula for net worth:

Net worth = Total assets − Total liabilities

Let's substitute the given values into the formula:

$15,000 = Value of Jewelry + \$4,000 + (−\$1,500) + \$8,500 + (−\$25,000) + \$20,000$
$+ (−\$7,500)$

Simplifying the equation, we get:

$15,000 = Value of Jewelry − \$1,500$

Adding $1,500 from both sides, we get:

$16,500 = Value of Jewelry$

Therefore, the current value of Jack's jewelry is $16,500.

4) Choice D is correct.

Let's first find the length of side QR: Side $QR = 1.1 \times PQ = 1.1 \times 18 = 19.8$ inches Then, we can find the length of side PR: Side $PR = \left(\frac{2}{3}\right) \times QR = \left(\frac{2}{3}\right) \times 19.8 = 13.2$ inches Now, we need to convert the lengths to feet, using the scale factor of $\frac{1}{8}$:

$$PQ = \frac{18}{8} = 2.25 \; feet$$

$$QR = \frac{19.8}{8} = 2.475 \; feet$$

$$PR = \frac{13.2}{8} = 1.65 \; feet$$

Finally, we can calculate the perimeter of the actual triangular face: Perimeter $= PQ + QR + PR = 2.25 + 2.475 + 1.65 = 6.375 \; feet$ Therefore, the correct answer is D, 6.375.

5) The answer is 20.

To find the total distance Joe covered during the given time, we need to calculate the distance covered during each leg of the journey and add them up.

Distance covered during the first leg (cycling for 2 hours at $10 \frac{km}{h}$):

Distance = Speed × Time

$Distance = 10 \frac{km}{h} \times 2h = 20 \; km$

Distance covered during the second leg (walking for 1 hour at $4\frac{km}{h}$):

$Distance = 4\frac{km}{h} \times 1\,h = 4\,km$

Distance covered during the third leg (cycling for 1.5 hours at $12\frac{km}{h}$):

$Distance = 12\frac{km}{h} \times 1.5\,h = 18\,km$

Total distance covered = Distance of first leg + Distance of second leg + Distance of third leg

Total distance covered = $20\,km + 4\,km + 18\,km = 42\,km$

Therefore, Joe covered a total distance of $42\,km$ during this time.

6) Choice B is correct.

There are four arrows and each one equals to $5 \rightarrow 4.\,(-5)$

7) Choice B is correct.

To find the individual with the highest weight-to-height ratio, we need to calculate the weight-to-height ratio for each individual. This can be done by dividing the weight of each individual (in pounds) by their height (in inches), and then comparing the ratios.

Calculating the weight-to-height ratios for each individual:

A: $\frac{170}{68} = 2.5$

B: $\frac{200}{72} = 2.78$

C: $\frac{150}{65} = 2.31$

D: $\frac{180}{70} = 2.57$

Therefore, individual B has the highest weight-to-height ratio of 2.78. So, the answer is B, Individual B.

8) Choice D is correct.

To calculate the tip amount, multiply the total bill by the tip rate as a decimal:

Tip amount = Total bill × Tip rate

Tip amount = $\$75 \times 0.15$

Tip amount = $\$11.25$

Therefore, the answer is D, $11.25.

9) Choice C is correct.

According to the graph, 32% of teenagers are interested to use Instagram.

Find 32% of 360°.

32% of 360° = 0.32 × 360° = 115.2°

10) Choice B is correct.

A $\frac{3}{4}$ probability of rain on Friday means that there is a 75% chance of precipitation occurring on that day. This is a relatively high percentage, which indicates that rainfall is quite likely.

In this case, option B, is the most accurate description of the situation, as it states that rainfall on Friday is highly probable. A 75% chance of rain is not a certainty, but it is a strong likelihood, making it more appropriate to describe the situation as highly probable rather than moderate (Option C), very unlikely (Option D), or stating that there is no chance of rain (Option A).

11) Choice D is correct.

To achieve at the most reliable conclusion, the university should use a sample that is representative of the entire population. Option D, 100 students from each school, would be the best sample size because it includes students from each school and ensures that the sample is large enough to provide reliable data.

12) The answer is 40.

To solve the problem, we can set up a proportion:

$$\frac{3}{5} = \frac{24}{x}$$

where x is the number of cheese pizzas.

To solve for x, we can cross-multiply:

$3x = 120 \rightarrow x = 40$

$x = 40$Therefore, there are 40 cheese pizzas in the pizza parlor.

The answer is C, 40.

13) Choice A is correct.

To find out which equation is true when $x = 3$, we substitute 3 for x in each equation and simplify.

A. $2x + 3 = 9$

Substituting $x = 3$: $2(3) + 3 = 6 + 3 = 9$

This equation is true when $x = 3$.

B. $x + 4 = 8$

Substituting $x = 3$: $3 + 4 = 7, 7 \neq 8$

This equation is false when $x = 3$.

C. $5x - 2 = 11$

Substituting $x = 3$: $5(3) - 2 = 15 - 2 = 13, 13 \neq 11$

This equation is false when $x = 3$.

D. $3x - 7 = 0$

Substituting $x = 3$: $3(3) - 7 = 9 - 7 = 2, 0 \neq 2$

This equation is false when $x = 3$.

Therefore, the equation that holds true when $x = 3$ is A. $2x + 3 = 9$.

14) Choice C is correct.

The volume of the tank is $3 \times 2 \times 2 = 12$ cubic feet. To find the amount of water needed to fill the tank to 60% capacity, we multiply the volume of the tank by 0.6, giving us 7.2 cubic feet. Finally, we multiply this result by 7.48 to convert cubic feet to gallons, giving us 53.86 gallons to the nearest tenth.

15) Choice B is correct.

If $\frac{5}{8}$ of a swimming pool can hold 2,400 gallons of water, then the total capacity of the swimming pool is:

$$\frac{2,400}{\frac{5}{8}} = 2,400 \times \left(\frac{8}{5}\right) = 3,840 \; gallons$$

Therefore, the total capacity of three such swimming pools is:

$$3 \times 3,840 = 11,520 \; gallons$$

16) Choice A is correct.

To find the range of the data, we need to subtract the smallest value from the largest value. In this case, the smallest value is 4 and the largest value is 9, so the range is:

Range = Largest value − Smallest value

$= 9 - 4 = 5$

To find the median of the data, we need to first put the data in order from smallest to largest:

$4, 5, 5, 6, 7, 8, 9$

The median is the middle value in this ordered set of data. Since there are 7 data points, the median is the $4th$ values:

$Median = 6$

Therefore, the answer is A, Range: 5; Median: 6.

17) Choice A is correct.

The inequality that represents all possible values of a and b, the number of each type of toy that the toy manufacturer can produce with the given budget of $3,000 is:

A. $50a + 30b \leq 3,000$

This inequality ensures that the cost of producing type A toys plus the cost of producing type B toys does not exceed the budget of $3,000.

18) Choice B is correct.

Solving for the radius using the formula for the circumference of a circle, we get $r = \frac{C}{2\pi} = \frac{18.84}{2\pi} \approx$ 3. We can then use the formula for the area of a circle, $A = \pi r^2$, to find the area. Substituting the value for r, we get $A = 3.14 \times 3^2 = 28.26$ square feet.

19) Choice D is correct.

We can use proportions to estimate the number of packages with a missing item. The proportion of packages with a missing item in the sample is $\frac{8}{150}$. We can use this proportion to estimate the number of packages with a missing item in the population as follows:

Number of packages with a missing item $= \frac{8}{150} \times 3,000 = 160$

We can predict that 160 packages out of the 3,000 will have a missing item.

20) The answer is $\frac{1}{3}$.

There are a total of $4 + 3 + 2 = 9$ balls in the box. The probability of selecting a green ball is the number of green balls divided by the total number of balls. Therefore, the probability is $\frac{3}{9}$, which can be simplified to $\frac{1}{3}$.

So, the answer is A, $\frac{1}{3}$.

21) Choice A is correct.

In an arithmetic sequence, each term is found by adding the common difference to the previous term. Since the common difference in this case is 3, we can find the next terms by adding 3 to the previous terms.

Starting with the first term of 7, we can add 3 to get the second term of 10, then add 3 again to get the third term of 13, and so on.

Therefore, the sequence $7, 10, 13, 16, 19, \ldots$ is an arithmetic progression with a common difference of 3 and a first term of 7.

22) Choice D is correct.

Out of the 100 people surveyed, 45 preferred burgers and 25 preferred fries. Therefore, the probability of someone preferring burgers is $\frac{45}{100} = 0.45$, and the probability of someone preferring fries is $\frac{25}{100} = 0.25$.

To compare the likelihood of someone preferring burgers to fries, we can calculate the ratio of their probabilities, $\frac{0.45}{0.25} = 1.8$. This means that someone is 1.8 times as likely to prefer burgers as fries.

Option A, which says that the person is twice as likely to prefer a burger as fries, is not correct. Option B, which is also not correct as the ratio of the probabilities of preferring pizza to chicken is $\frac{10}{20} = 0.5$, is not equal to 4. Option C, which is not correct either as the probability of preferring chicken or pizza is $0.2 + 0.1 = 0.3$, is greater than the probability of preferring fries (0.25). Option D, which is correct as the probability of preferring burger or chicken is $0.45 + 0.2 = 0.65$, is greater than the probability of preferring fries or pizza ($0.25 + 0.1 = 0.35$).

23) Choice A is correct.

According to the given equation, the velocity of the car increases by 5 meters per second for every second that passes, starting from an initial velocity of 2 meters per second when $t = 0$. Therefore, the car is accelerating at a rate of 5 meters per second per second.

24) Choice B is correct.

In a triangle, the three interior angles always add up to 180 degrees.

Therefore: $(3x) + (7x + 12) + (6x + 8) = 180 \rightarrow 16x + 20 = 180 \rightarrow 16x = 160 \rightarrow x = 10$

25) Choice A is correct.

To find the length of the dress in the sketch, we need to convert the actual length of the dress from inches to centimeters using the given scale.

1 centimeter represents 2 inches, which means that:

$$1 \ inch = \frac{1}{2} \ centimeter$$

$$60 \ inches = 60 \times \frac{1}{2} \ centimeters$$

$$60 \ inches = 30 \ centimeters$$

Therefore, the length of the dress in the sketch is 30 cm (Option A).

26) Choice C is correct.

The average speed of the airplane is $\frac{1,600}{4} = 400 \frac{km}{h}$.

The average speed of the helicopter is $\frac{800}{2} = 400 \frac{km}{h}$.

Therefore, the ratio of the average speed of the airplane to the average speed of the helicopter is $1:1$, which is option C.

27) Choice D is correct.

The statistical measure that best describes the variability of the sales data is the Range (Option D). The range is the difference between the largest and smallest values in a dataset. In this case, the largest value is 35 and the smallest value is 15, so the range would be 20.

28) Choice C is correct.

$-18 < 5x + 7 < 12 \rightarrow$ Subtract 7 from each side. $-18 - 7 < 5x + 7 - 7 < 12 - 7$

$\rightarrow -25 < 5x < 5 \rightarrow$ Divide all sides by 5.

$\frac{-25}{-5} < \frac{5x}{5} < \frac{5}{5} \rightarrow -5 < x < 1$

29) Choice A is correct.

To solve this problem, we first need to find out how many cell phones and laptops Mr. Rodriguez sells per day:

Cell phones: $\frac{51}{3} = 17$ per day

Laptops: $\frac{102}{3} = 34$ per day

Now we can use this information to find out how many cell phones and laptops Mr. Rodriguez will sell in 9 days:

Cell phones: $17 \times 9 = 153$

Laptops: $34 \times 9 = 306$

Therefore, Mr. Rodriguez will sell 153 cell phones and 306 laptops in 9 days. The difference between the number of laptops and cell phones sold is $306 - 153 = 153$. Therefore, the answer is option A, 153.

30) Choice A is correct.

To solve the problem, we can use the formula for direct variation:

$y = kx^4$

where k is the constant of proportionality. We can find k by substituting the given values:

$2 = k(2)^4 \rightarrow k = \frac{1}{8}$

Now we can use this value of k to find x when $y = 512$:

$512 = \left(\frac{1}{8}\right)x^4 \rightarrow x^4 = 4,096 \rightarrow x = 8$

Therefore, when $y = 512, x = 8$.

31) The answer is 110.

To find the amount of wood needed to make one chair, we need to divide the total amount of wood by the number of chairs.

$Amount\ of\ wood\ needed\ for\ one\ chair = \frac{Total\ amount\ of\ wood}{Number\ of\ chairs}$

$Amount\ of\ wood\ needed\ for\ one\ chair = \frac{5,500\ cm^2}{50\ chairs}$

$Amount\ of\ wood\ needed\ for\ one\ chair = 110\ cm^2$

Therefore, the answer is option B, $110\ cm^2$.

32) Choice C is correct.

To find out the minimum score that David must obtain on his fifth and final test to qualify for the advanced placement program, we can use the formula:

$Average = \frac{Total\ score\ on\ all\ tests}{Number\ of\ tests}$

We know that David needs to achieve an average of 80% in his physics class to qualify for the advanced placement program. Therefore, we can write:

$80\% = \frac{78\%+81\%+85\%+88\%+x\%}{5}$

where $x\%$ is the score that David needs to achieve on his fifth and final test.

Simplifying the equation, we get: $80\% = \frac{332\%+x\%}{5}$

Multiplying both sides by 5, we get: $400\% = 332\% + x\%$

Subtracting 332% from both sides, we get: $68\% = x\%$

Therefore, David needs to achieve a minimum score of 68% on his fifth and final test to qualify for the advanced placement program.

33) Choice D is correct.

John has not completed 7 out of 25 problems in his physics homework.

To find the percentage of problems John has not completed, we can use the formula:

$Percentage = \left(\frac{number\ of\ problems\ not\ completed}{total\ number\ of\ problems}\right) \times 100\%$

So, in this case, the percentage of problems John has not completed is: $\left(\frac{7}{25}\right) \times 100\% = 28\%$

Therefore, the answer is option D, 28%.

34) Choice D is correct.

To solve the system of equations, we can use the method of substitution or elimination. Here, we will use the method of elimination to find the values of x and y that satisfy both equations.

Multiplying the first equation by 2, we get: $6x - 4y = 12$

Adding the second equation, we get: $11x = 22$

Therefore, $x = 2$.

Substituting $x = 2$ in the first equation, we get:

$3(2) - 2y = 6 \rightarrow 6 - 2y = 6 \rightarrow -2y = 0 \rightarrow y = 0$

So, the solution is $(x, y) = (2,0)$.

35) Choice A is correct.

The volume of a cylinder $= \pi r^2 h$, the radius of the cylinder is 4 meters and its height is 15 meters. Therefore, the volume of a cylinder is $\pi(4)^2(15) = 753.6$.

36) Choice B is correct.

The total value of Sarah's assets and liabilities can be calculated as follows:

$Total\ Assets - Total\ Liabilities = Net\ Worth$

$\$360{,}000 - \$215{,}500 = \$144{,}500$

Therefore, the answer is B, $\$144{,}500$.

37) Choice A is correct.

Find the corresponding sides and write a proportion: $\frac{AB}{BC} = \frac{EF}{FG}$

Substitute 22 for AB, 21 for BC, and 14 for EF. Then: $\frac{22}{21} = \frac{14}{FG}$

38) Choice C is correct.

We can start by converting the mixed number $3\frac{1}{5}$ to an improper fraction:

$3\frac{1}{5} = \frac{3 \times 5 + 1}{5} = \frac{16}{5}$

Then we can divide 22.68 by $\frac{16}{5}$:

$22.68 \div \left(\frac{16}{5}\right) = 22.68 \times \frac{5}{16} = 7.09$

Therefore, the answer is C, 7.09.

39) Choice D is correct.

Let x equal the smallest angle of the triangle. Then, the three angles are $3x$, $4x$, and $8x$. The sum of the angles of a triangle is 180. Set up an equation using this to find x:

$3x + 4x + 8x = 180 \rightarrow 15x = 180 \rightarrow x = 12$

Since the question asks for the measure of the largest angle, $8x = 8(12) = 96°$.

40) Choice B is correct.

To find the probability of rolling a total of 9 or 12, we need to first determine the number of ways we can get each of these totals.

For a total of 9, the possible combinations are $(3,6), (4,5), (5,4)$, and $(6,3)$, which gives us a total of 4 possible outcomes.

For a total of 12, the only possible combination is $(6,6)$, which gives us 1 possible outcome.

Therefore, the total number of outcomes that result in a total of 9 or 12 is $4 + 1 = 5$.

Since there are 36 possible outcomes when rolling two dice (6 possible outcomes for each of the 6 faces of a die), the probability of rolling a total of 9 or 12 is $\frac{5}{36}$.

STAAR Mathematics Practice Test 6
Answers and Explanations

1)   Choice A is correct.

Area of the circle is less than 49π. Use the formula of areas of circles.

$Area = \pi r^2 \rightarrow 49\pi > \pi r^2 \rightarrow 49 > r^2 \rightarrow r < 7$

Radius of the circle is less than 7. Let's put 7 for the radius. Now, use the circumference formula: $Circumference = 2\pi r = 2\pi(7) = 14\pi$. Since the radius of the circle is less than 7. Then, the circumference of the circle must be less than 14π. Only choice A is less than 14π.

2)   Choice C is correct.

Use the formula for Percent of Change: $\frac{New\ Value - Old\ Value}{Old\ Value} \times 100\%$

$\frac{27-45}{45} \times 100\% = -40\%$ (Negative sign here means that the new price is less than old price).

3)   Choice B is correct.

The question is this: 1.75 is what percent of 1.40? Use percent formula:

$part = \frac{percent}{100} \times whole \rightarrow 1.75 = \frac{percent}{100} \times 1.40 \rightarrow 1.75 = \frac{percent \times 1.40}{100} \rightarrow 175 = percent \times 1.40 \rightarrow percent = \frac{175}{1.40} = 125$

4)   Choice A is correct.

The percent of girls playing tennis is: $50\% \times 20\% = 0.50 \times 0.20 = 0.1 = 10\%$

5)   Choice B is correct.

Write the numbers in order: $3, 5, 7, 8, 13, 15, 18$. Since we have 7 numbers (7 is odd), then the median is the number in the middle, which is 8

6)   Choice C is correct.

Three times of 24,000 is 72,000. One sixth of them cancelled their tickets.

One sixth of 72,000 equals $12,000 (\frac{1}{6} \times 72,000 = 12,000)$.

$60,000 (72,000 - 12,000 = 60,000)$ fans are attending this week.

7)   Choice C is correct.

The average speed of john is: $140 \div 7 = 20 \ km$, The average speed of Alice is: $180 \div 4 = 45 \ km$, Write the ratio and simplify. $20 : 45 = 4 : 9$

8)   Choice C is correct.

Plug in the value of x and y. $3(x - 2y) + (2 - x)^2$ when $x = 5$ and $y = -3$

$3(x - 2y) + (2 - x)^2 = 3(5 - 2(-3)) + (2 - 5)^2 = 3(5 + 6) + (-3)^2 = 33 + 9 = 42$

9)   Choice D is correct.

Isolate and solve for x. $\frac{2}{3}x + \frac{1}{6} = \frac{1}{3} \rightarrow \frac{2}{3}x = \frac{1}{3} - \frac{1}{6} = \frac{1}{6} \rightarrow \frac{2}{3}x = \frac{1}{6}$

Multiply both sides by the reciprocal of the coefficient of x. $(\frac{3}{2})\frac{2}{3}x = \frac{1}{6}(\frac{3}{2}) \rightarrow x = \frac{3}{12} = \frac{1}{4}$

10)   Choice B is correct.

Use the information provided in the question to draw the shape.

Use Pythagorean Theorem: $a^2 + b^2 = c^2$

$12^2 + 16^2 = c^2 \rightarrow 144 + 256 = c^2 \rightarrow 400 = c^2 \rightarrow c = 20$

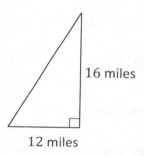
16 miles

12 miles

11)   Choice B is correct.

The question is this: 600 is what percent of 640? Use percent formula:

$part = \frac{percent}{100} \times whole \rightarrow 600 = \frac{percent}{100} \times 640 \rightarrow 600 = \frac{percent \times 640}{100}$

$$\rightarrow 60{,}000 = percent \times 640 \rightarrow percent = \frac{60{,}000}{640} = 93.75$$

600 is 93.75% of 640. Therefore, the discount is: $100\% - 93.75\% = 6.25\%$

12)   Choice B is correct.

If the score of Mia is 40, therefore the score of Ava is 20. Since, the score of Emma is half of Ava, therefore, the score of Emma is 10.

13)   Choice D is correct.

If 19 balls are removed from the bag at random, there will be one ball in the bag. The probability of choosing a brown ball is 1 out of 20. Therefore, the probability of not choosing a brown ball is 19 out of 20 and the probability of not having a brown ball after removing 19 balls is the same.

14) Choice C is correct.

The weight of 15.2 meters of the rope is, $15.2 \times 600 \ g = 9,120 \ g$. and $1 \ kg = 1,000 \ g$, therefore, $9,120 \ g \div 1,000 = 9.12 \ kg$.

15) Choice B is correct.

8% of the volume of the solution is alcohol. Let x be the volume of the solution.
Then: $8\% \ of \ x = 38.4 \ ml \rightarrow 0.08x = 38.4 \rightarrow x = 38.4 \div 0.08 = 480 \ ml$

16) Choice D is correct.

First calculate the number of feet that 1 inch represents: $100 \ ft \div 5 \ in = 20 \frac{ft}{in}$

Then multiply this by the total number of inches: $18 \ in \times 20 \frac{ft}{in} = 360 \ ft$

17) Choice D is correct.

$\frac{2}{5} \times 25 = \frac{50}{5} = 10$

18) Choice D is correct.

Simplify. $6x^2y^3(2x^2y)^3 = 6x^2y^3(8x^6y^3) = 48x^8y^6$

19) Choice D is correct.

Use Pythagorean Theorem: $a^2 + b^2 = c^2$
$9^2 + 12^2 = c^2 \rightarrow 81 + 144 = c^2 \rightarrow 225 = c^2 \rightarrow c = 15$

20) Choice C is correct.

To find the discount, multiply the number by $(100\% - rate \ of \ discount)$.
Therefore, for the first discount we get: $(D)(100\% - 15\%) = (D)(0.85) = 0.85D$
For increase of 10%: $(0.85D)(100\% + 10\%) = (0.85D)(1.10) = 0.935 \ D = 93.5\% \ of \ D$

21) The answer is − 64.

Use PEMDAS (order of operation):

$[3 \times (-14) - 48] - (-14) + [3 \times 8] \div 2 = [-42 - 48] + 14 + 24 \div 2 = -90 + 14 + 12 = -64$

22) Choice D is correct.

Let x be the original price. If the price of a laptop is decreased by 10% to $450, then:

$90\% \ of \ x = 450 \rightarrow 0.90x = 450 \rightarrow x = 450 \div 0.90 = 500$

23) The answer is 52.

$average = \dfrac{sum \ of \ terms}{number \ of \ terms} \rightarrow 25 = \dfrac{13+15+20+x}{4} \rightarrow 100 = 48 + x \rightarrow x = 52$

24) Choice D is correct.

The ratio of boy to girls is $4 : 7$. Therefore, there are 4 boys out of 11 students. To find the answer, first divide the total number of students by 11, then multiply the result by 4.

$55 \div 11 = 5 \Rightarrow 5 \times 4 = 20$. There are 20 boys and 35 $(55 - 20)$ girls. So, 15 more boys should be enrolled to make the ratio $1 : 1$.

25) Choice C is correct.

Let x be the number of soft drinks for 160 guests. It's needed to have a proportional ratio to find x. $\dfrac{14 \ soft \ drinks}{16 \ guests} = \dfrac{x}{160 \ guests} \rightarrow x = \dfrac{160 \times 14}{16} \Rightarrow x = 140$

26) Choice D is correct.

Let x be all expenses, then $\dfrac{22}{100}x = \$660 \rightarrow x = \dfrac{100 \times \$660}{22} = \$3,000$.

He spent for his rent: $\dfrac{27}{100} \times \$3,000 = \$810$

27) Choice C is correct.

Use the percentage formula:

$part = \dfrac{percent}{100} \times whole \rightarrow 35 = \dfrac{percent}{100} \times 20 \rightarrow 35 = \dfrac{percent \times 20}{100} \rightarrow 35 = \dfrac{percent \times 2}{10}$

Then, multiply both sides by 10. $350 = percent \times 2$, and divide both sides by 2:

$175 = percent$

28) Choice C is correct.

Let x be the number. Write the equation and solve for x. $\frac{2}{3} \times 18 = \frac{2}{5} \times x \rightarrow \frac{2 \times 18}{3} = \frac{2x}{5}$, use cross multiplication to solve for x. $5 \times 36 = 2x \times 3 \rightarrow 180 = 6x \rightarrow x = 30$

29) Choice D is correct.

If the length of the box is 36, then the width of the box is one third of it, 12, and the height of the box is 4 (one third of the width). The volume of the box is: $V = lwh = (36)(12)(4) = 1,728$

30) The answer is 80.

The fail rate is 11 out of $55 = \frac{11}{55}$. Change the fraction to percent: $\frac{11}{55} \times 100\% = 20\%$

20 percent of students failed. Therefore, 80 percent of students passed the exam.

31) Choice A is correct.

Let L be the price of the laptop and C be the price of the computer. $2(L) = 3(C)$ and $L = \$100 + C$

Therefore, $2(\$100 + C) = 3C \rightarrow \$200 + 2C = 3C \rightarrow C = \$200$

32) The answer is 12.

The width of the rectangle is twice its length. Let x be the length. Then, $width = 2x$
The perimeter of the rectangle is $2(width + length) = 2(2x + x) = 72 \Rightarrow 6x = 72 \rightarrow x = 12$. The length of the rectangle is 12 meters.

33) Choice D is correct.

The distance between Bob and Mike is 12 miles. Bob is running at 6.5 miles per hour and Mike is running at the speed of 8 miles per hour. Therefore, every hour the distance is reduced by 1.5 miles. $12 \div 1.5 = 8$

34) Choice D is correct.

11 out of 44 or $\frac{11}{44}$ marbles are red. Convert the fraction to percent: $\frac{11}{44} \times 100\% = 25\%$
25 percent of marbles are red. Therefore, 75 percent of marbles are NOT red.

35) The answer is 61.2.

$The\ average = \frac{sum\ of\ terms}{number\ of\ terms}$. The sum of the weight of all girls is: $20 \times 60 = 1,200\ kg$, The sum of the weight of all boys is: $30 \times 62 = 1,860\ kg$. The sum of the weight of all students is: $1,200 + 1,860 = 3,060\ kg$.

$average = \frac{3,060}{50} = 61.2$

36) Choice C is correct.

Use the simple interest formula: $I = prt$ ($I = interest, p = principal, r = rate, t = time$)
$I = (9,000)(0.045)(5) = 2,025$

37) Choice D is correct.

Let x be the integer. Then: $2x - 6 = 68$. Add 6 to both sides: $2x = 74$. Divide both sides by 2: $x = 37$

38) Choice B is correct.

Plug in each pair of numbers in the equation: $3x + 5y = 7$
A. $(2, 1)$: $3(2) + 5(1) = 11$
B. $(-1, 2)$: $3(-1) + 5(2) = 7$
C. $(-2, 2)$: $3(-2) + 5(2) = 4$
D. $(2, 2)$: $3(2) + 5(2) = 16$
Choice B is correct.

39) The answer is 175.

To find the number of possible outfit combinations, multiply the number of options for each item:

$5 \times 7 \times 5 = 175$

40) Choice C is correct.

The sum of supplementary angles is 180. Let x be that angle. Therefore, $x + 4x = 180$
$5x = 180$. Divide both sides by 5: $x = 36$

STAAR Mathematics Practice Test 7
Answers and Explanations

1) Choice B is correct.

For sum of 5: (1&4) and (4&1), (2&3) and (3&2), therefore we have 4 options.

For sum of 8: (5&3) and (3&5), (4&4) and (2&6), (6&2), we have 5 options. To get a sum of 5 or 8 for two dice: $4 + 5 = 9$. Since, we have $6 \times 6 = 36$ total number of options, the probability of getting a sum of 5 and 8 is 9 out of 36 or $\frac{9}{36} = \frac{1}{4}$

2) The answer is 10.

Use formula of rectangle prism volume. $V = (length)(width)(height) \rightarrow$
$$2,500 = (25)(10)(height) \Rightarrow height = 2,500 \div 250 = 10$$

3) Choice D is correct.

Solve for x. $-4 \le 4x - 8 < 1 \rightarrow 6$ (Add 8 all sides) $-4 + 8 < 4x - 8 + 8 < 16 + 8 \rightarrow 4 < 4x < 2 \rightarrow 4$ (divide all sides by 4) $1 \le x < 6$

x is between 1 and 6. Choice D represents this inequality.

4) Choice B is correct.

Since 0.0099 is equal to 0.99%, the closest to that value is 0.1%.

5)   Choice B is correct.

The probability of choosing a Club is $\frac{13}{52} = \frac{1}{4}$

6)   Choice C is correct.

Use distance formula: $Distance = Rate \times time \rightarrow 380 = 40 \times T$, divide both sides by 40. $\frac{380}{40} = T \rightarrow T = 9.5$ hours. Change hours to minutes for the decimal part. 0.5 hours $= 0.5 \times 60 = 30$ minutes

7)   Choice C is correct.

$(16 \times 36) + (9 \times 12) + 10 = 694$

8)   Choice D is correct.

To find the discount, multiply the number by $(100\% - rate\ of\ discount)$.

Therefore, for the first discount we get: $(500)(100\% - 25\%) = (500)(0.75)$

For the next 15% discount: $(500)(0.75)(0.85)$

9)   Choices C is correct.

y is the intersection of the three circles. Therefore, it must be odd (From circle A), negative (From circle B), and multiple of 5 (From circle C).

From the choices provided, only -5 is odd, negative and multiple of 5.

10)   The answer is 5.

Set up a proportion to solve: $\dfrac{\frac{5}{14}\ cherry}{\frac{5}{70}\ apple} = \dfrac{x\ cherry}{1\ apple} \rightarrow \dfrac{5}{14} \times \dfrac{70}{5} = x \rightarrow x = \dfrac{70}{14} = \dfrac{10}{2} \rightarrow x = 5$

11)   Choice C is correct.

Use this formula: Percent of Change: $\dfrac{New\ Value - Old\ Value}{Old\ Value} \times 100\%$

$\dfrac{28{,}000 - 18{,}200}{28{,}000} \times 100\% = -35\%$.

The negative sign means that the price decreased.

12)   Choice D is correct.

Plug in 140 for F and then solve for C. $C = \dfrac{5}{9}(F - 32) \rightarrow C = \dfrac{5}{9}(140 - 32) \rightarrow$

$C = \dfrac{5}{9}(108) = 60$

13)   Choice C is correct.

The square of a number is $\dfrac{36}{64}$, then the number is the square root of $\dfrac{36}{64}$: $\sqrt{\dfrac{36}{64}} = \dfrac{6}{8}$

The cube of the number is: $\left(\dfrac{6}{8}\right)^3 = \dfrac{216}{512}$

14) Choice C is correct.

Surface Area of a cylinder $= 2\pi r(r + h)$, the radius of the cylinder is 2, $(4 \div 2)$ inches and its height is 8 inches. Therefore, Surface Area of a cylinder $= 2\pi(2)(2 + 8) = 40\pi$

15) Choice D is correct.

$3x - 5 = 8.5 \rightarrow 3x = 8.5 + 5 = 13.5 \rightarrow x = \frac{13.5}{3} = 4.5$

Then; $6x + 3 = 6(4.5) + 3 = 27 + 3 = 30$

16) Choice D is correct.

$Average = \frac{sum\ of\ terms}{number\ of\ terms}$

$\rightarrow$ (Average of 6 numbers) $12 = \frac{sum\ of\ numbers}{6} \rightarrow$ sum of 6 numbers is $12 \times 6 = 72$.

$\rightarrow$ (Average of 4 numbers) $10 = \frac{sum\ of\ numbers}{4} \rightarrow$ sum of 4 numbers is $10 \times 4 = 40$.

$sum\ of\ 6\ numbers - sum\ of\ 4\ numbers = sum\ of\ 2\ numbers \rightarrow 72 - 40 = 32$.

Average of 2 numbers is, $\frac{32}{2} = 16$.

17) Choice C is correct.

$Probability = \frac{number\ of\ desired\ outcomes}{number\ of\ total\ outcomes} = \frac{10}{15+10+10+25} = \frac{10}{60} = \frac{1}{6}$

18) Choice D is correct.

Change the numbers to decimal and then compare. $\frac{11}{15} = 0.73 \ldots, 0.74, 75\% = 0.75, \frac{19}{25} = 0.76$

Therefore: $\frac{11}{15} < 0.74 < 75\% < \frac{19}{25}$

19) Choice C is correct.

To find the number of possible outfit combinations, multiply number of options for each factor: $5 \times 7 \times 3 = 105$

20) Choice C is correct.

$4 \div \frac{1}{3} = 12$

21) Choice A is correct.

The diagonal of the square is 4. Let x be the side.

Use Pythagorean Theorem:

$a^2 + b^2 = c^2 \rightarrow x^2 + x^2 = 4^2 \rightarrow 2x^2 = 4^2 \rightarrow 2x^2 = 16 \rightarrow x^2 = 8 \rightarrow x = \sqrt{8}$

The area of the square is: $\sqrt{8} \times \sqrt{8} = 8$

22) Choice C is correct.

The ratio of boy to girls is $2:3$. Therefore, there are 2 boys out of 5 students. To find the answer, first divide the total number of students by 5, then multiply the result by 2.

$600 \div 5 = 120 \rightarrow 120 \times 2 = 240$

23) Choice B is correct.

Let x be the width of the rectangle. Use Pythagorean Theorem:

$a^2 + b^2 = c^2 \rightarrow x^2 + 5^2 = 13^2 \rightarrow x^2 + 25 = 169 \rightarrow x^2 = 169 - 25 = 144 \rightarrow x = 12$

Area of the rectangle $= length \times width = 5 \times 12 = 60$

24) Choice A is correct.

Mr. Jones saves 5,000 out of 85,000 which equals to $\frac{5,000}{85,000} = \frac{5}{85} = \frac{1}{17}$

25) Choice C is correct.

Let x be the number. Write the equation and solve for x. $(32 - x) \div x = 3$.

Multiply both sides by x. $(32 - x) = 3x$, then add x to both sides. $32 = 4x$, now divide both sides by 4. $x = 8$

26) Choice D is correct.

$Volume\ of\ a\ box = length \times width \times height = 8\ cm \times 4\ cm \times 5\ cm = 160\ cm^3$

27) Choice B is correct.

The population is increased by 10% and 20%. 10% increase changes the population to 110% of the original population. For the second increase, multiply the result by 120%.

$(1.10) \times (1.20) = 1.32 = 132\%$. 32 percent of the population is increased after two years.

28) The answer is 20.

Recall that the formula for the average is: $Average = \frac{sum\ of\ data}{number\ of\ data}$

First, compute the total weight of all balls in the basket: $25\ g = \frac{total\ weight}{20\ balls}$

$25\ g \times 20 = total\ weight \rightarrow total\ weight = 500\ g$

Next, find the total weight of the 5 largest marbles:

$40\ g = \frac{total\ weight}{5\ marbles} \rightarrow 40\ g \times 5 = total\ weight \rightarrow total\ weight = 200\ g$

The total weight of the heaviest balls is $200g$. Then, the total weight of the remaining 15 balls is: $300\ g$ ($500\ g - 200\ g = 300\ g$).

The average weight of the remaining balls: $Average = \frac{300\ g}{15\ marbles} = 20\ g$ per ball

29) Choice C is correct.

The area of the floor is: $5\ cm \times 20\ cm = 100\ cm^2$, The number of tiles needed:

$100 \div 4 = 25$

30) Choices B is correct.

The sum of all the internal angles of a simple polygon is $180(n-2)$ where n is the number of sides, so $180(4-2) = 180 \times 2 = 360$. Vertical angles are congruent.

Then: $360 = 120 + 120 + 98 + c \rightarrow c = 360 - 338 = 22$

31) Choice B is correct.

The equation of a line in slope-intercept form is: $y = mx + b$.

Solve for y. $3x - y = 6 \rightarrow -y = -3x + 6$.

Divide both sides by (-1). Then, $-y = -3x + 6 \rightarrow y = 3x - 6$.

The slope of this line is 3. The product of the slopes of two perpendicular lines is -1.

Therefore, the slope of a line that is perpendicular to this line is:

$m_1 \times m_2 = -1 \Rightarrow 3 \times m_2 = -1 \Rightarrow m_2 = \frac{-1}{3} = -\frac{1}{3}$

32) Choice B is correct.

The sum of 8 numbers is greater than 240 and less than 320. Then, the average of the 8 numbers must be greater than 30 and less than 40.

$$\frac{240}{8} < x < \frac{320}{8} \rightarrow 30 < x < 40$$

The only choice that is between 30 and 40 is 35.

33) Choice B is correct.

$$average\ (mean) = \frac{sum\ of\ terms}{number\ of\ terms} \rightarrow 60 = \frac{sum\ of\ terms}{40} \rightarrow sum = 60 \times 40 = 2,400$$

The difference of 74 and 94 is 20. Therefore, 20 should be subtracted from the sum.

$2,400 - 20 = 2,380.$

$$mean = \frac{sum\ of\ terms}{number\ of\ terms} \Rightarrow mean = \frac{2,380}{40} = 59.5$$

34) Choice B is correct.

Let L be the length of the rectangular and W be the width of the rectangular.

Then, $L = 4W + 3$

The perimeter of the rectangle is 36 meters. Therefore: $2L + 2W = 36 \rightarrow L + W = 18$

Replace the value of L from the first equation into the second equation and solve for W:

$(4W + 3) + W = 18 \rightarrow 5W + 3 = 18 \rightarrow 5W = 15 \rightarrow W = 3$

The width of the rectangle is 3 meters and its length is: $L = 4W + 3 = 4(3) + 3 = 15$

The area of the rectangle is: $Length \times Width = 3 \times 15 = 45$

35) Choice D is correct.

$$\frac{1}{3} \times 5\frac{1}{4} = \frac{1}{3} \times \frac{21}{4} = \frac{21}{12}$$

Converting $\frac{21}{12}$ to a mixed number gives: $\frac{21}{12} = 1\frac{9}{12} = 1\frac{3}{4}$

36) Choice D is correct.

$$2\frac{2}{3} - 1\frac{5}{6} = 2\frac{4}{6} - 1\frac{5}{6} = \frac{16}{6} - \frac{11}{6} = \frac{5}{6}$$

37) The answer is 90.

The perimeter of the trapezoid is $36cm$. Therefore, the missing side (height) is
$40 - 13 - 10 - 5 = 12$. Area of a trapezoid: $A = \frac{1}{2}h(b_1 + b_2) = \frac{1}{2}(12)(5 + 10) = 90$

38) Choice A is correct.

Let x be the number of years. Therefore, \$3,000 per year equals $3,000x$.

Starting from \$24,000 annual salary means you should add that amount to $3,000x$.

Income more than that is: $I > 3,000x + 24,000$

39) Choice D is correct.

Solve for x. $-1 \le 2x - 3 < 1 \to$ (Add 3 all sides) $-1 + 3 \le 2x - 3 + 3 < 1 + 3 \to$
$2 \le 2x < 4 \to$ (Divide all sides by 2) $1 \le x < 2$.

x is between 1 and 2. Choice D represents this inequality.

40) Choice C is correct.

x directly proportional to the square of y. Then: $x = cy^2 \to 12 = c(2)^2 \to 12 = 4c \to c = 3$.

The relationship between x and y is: $x = 3y^2 \to x = 75$

$75 = 3y^2 \to y^2 = \dfrac{75}{3} = 25 \to y = 5$

STAAR Mathematics Practice Test 8
Answers and Explanations

1)   Choice C is correct.

To find the area of the shaded region, find the difference of the area of two circles. (S_1: the area of bigger circle. S_2: the area of the smaller circle).

Use the area of a circle formula: $S = \pi r^2$

$S_1 - S_2 = \pi(5\ cm + 3\ cm)^2 - \pi(5\ cm)^2 \rightarrow S_1 - S_2 = 64\pi\ cm^2 - 25\pi\ cm^2 \rightarrow S_1 - S_2 = 39\pi\ cm^2$

2)   Choice D is correct.

William ate $\frac{1}{4}$ of 8 parts of his pizza that it means 2 parts out of 8 parts ($\frac{1}{4}$ of 8 parts $= x \rightarrow x = 2$) and left 6 parts. Ella ate $\frac{1}{2}$ of 8 parts of her pizza that it means 4 parts out of 8 parts ($\frac{1}{2}$ of 8 parts $= x \rightarrow x = 4$) and left 4 parts. Therefore, they ate $(4 + 2)$ parts out of $(8 + 8)$ parts of their pizza and left $(6 + 4)$ parts out of $(8 + 8)$ parts of their pizza. It means: $\frac{10}{16}$.

After simplification we have: $\frac{5}{8}$

3)   Choice C is correct.

Simplify: $\frac{48x^3y^8}{8x^2y^5} = 6xy^3$

4)   Choice D is correct.

Let's review the choices when $x = 1$.

A. $f(x) = x^2 - 5$, if $x = 1 \rightarrow f(1) = (1)^2 - 5 = 1 - 5 = -4 \neq 5$

B. $f(x) = x^2 - 1$, if $x = 1 \rightarrow f(1) = (1)^2 - 1 = 1 - 1 = 0 \neq 5$

C. $f(x) = \sqrt{x + 2}$, if $x = 1 \rightarrow f(1) = \sqrt{1 + 2} = \sqrt{3} \neq 5$

D. $f(x) = \sqrt{x} + 4$, if $x = 1 \rightarrow f(1) = \sqrt{1} + 4 = 5$

Only choice D provides a correct answer.

5)   Choice B is correct.

$12.9 \times 3.8 = 49.02\ L$

6) Choice D is correct.

Let x equal the smallest angle of the triangle. Then, the three angles are x, $3x$, and $5x$. The sum of the angles of a triangle is 180. Set up an equation using this to find x:

$x + 3x + 5x = 180 \rightarrow 9x = 180 \rightarrow x = 20$

Since the question asks for the measure of the largest angle, $5x = 5(20) = 100°$

7) Choice D is correct.

Find the corresponding sides and write a proportion: $\frac{AB}{BC} = \frac{EF}{FG}$

Substitute 12 for AB, 8 for BC, and 8 for EF. Then: $\frac{12}{8} = \frac{8}{FG}$

8) Choice B is correct.

To find the percentage of increased use this formula:

$\frac{new\ price - original\ price}{original\ price} \times 100 \rightarrow \frac{35-28}{28} \times 100 = \frac{7}{28} \times 100 = 25\%$

9) Choice B is correct.

Let x be the regular price. $0.15x = 6 \rightarrow x = \frac{6}{0.15} = 40$

10) Choice C is correct.

Let x to be the height of the triangle, then use this formula:

$A = \frac{b \times h}{2} \rightarrow 36 = \frac{4 \times x}{2} \rightarrow 4x = 72 \rightarrow x = 18$

11) Choice C is correct.

The general slope–intercept form of the equation of a line is $y = mx + b$, where m is the slope and b is the y −intercept.

Select two coordinate points: $A(2, 0)$ and $B(5, 3)$

Then find the slope: $= \frac{y_2 - y_1}{x_2 - x_1} = \frac{3-0}{5-2} = \frac{3}{3} = 1 \rightarrow m = 1$

b is the y −intercept, so $b = -2$

The equation of the line is: $y = x - 2$

12) The answer is -25.6.

$-8 \times 3.2 = -25.6$

13) Choice A is correct.

Write the ratio and solve for x: $\frac{45}{40} = \frac{2x+4}{16} \rightarrow 40(2x + 4) = 45 \times 16 \rightarrow 80x + 160 = 720 \rightarrow 80x = 560 \rightarrow x = 7$

14) Choice B is correct.

Only choice B is correct. Other choices don't work in the equation.

If $x = -8$:

Option A: $x(2x - 4) = 120 \rightarrow -8(2(-8) - 4) = 120 \rightarrow 160 \neq 120$, this is not true.

Option B: $8(4 - x) = 96 \rightarrow -8(4 - (-8)) = 96 \rightarrow 96 = 96$, this is true.

Option C: $2(4x + 6) = 79 \rightarrow 2(4(-8) + 6) = 79 \rightarrow 52 \neq 79$, this is not true.

Option D: $6x - 2 = -46 \rightarrow 6(-8) - 2 = -46 \rightarrow -50 \neq -46$, this is not true.

15) Choice A is correct.

Let x be the number of balls. Then: $\frac{1}{3}x + \frac{1}{6}x + \frac{1}{4}x + 12 = x$

$\left(\frac{1}{3} + \frac{1}{6} + \frac{1}{4}\right)x + 12 = x \rightarrow \left(\frac{9}{12}\right)x + 12 = x \rightarrow \frac{3}{4}x + 12 = x \rightarrow \frac{1}{4}x = 12 \rightarrow x = 48$

In the bag of small balls $\frac{1}{6}$ are white, then: $\frac{48}{6} = 8$. There are 8 white balls in the bag.

16) Choice B is correct.

Substitute numbers $-5, -1, 1, 2, 5, 8$ for x:

$x = -5 \rightarrow -3(-5) + 2 > 4 \rightarrow 17 > 4$, This is true.

$x = -1 \rightarrow -3(-1) + 2 > 4 \rightarrow 5 > 4$, This is true.

$x = 1 \rightarrow -3(1) + 2 > 4 \rightarrow -1 > 4$, This is not true.

$x = 2 \rightarrow -3(2) + 2 > 4 \rightarrow -4 > 4$, This is not true.

$x = 5 \rightarrow -3(5) + 2 > 4 \rightarrow -13 > 4$, This is not true.

$x = 8 \rightarrow -3(8) + 2 > 4 \rightarrow -22 > 4$, This is not true.

So, choice B is correct.

17) Choice D is correct.

$\frac{3}{8} = 0.375 = 37.50\%$

18) Choice B is correct.

$\alpha = 180° - 112° = 68°$

$\beta = 180° - 135° = 45°$

$x + \alpha + \beta = 180° \rightarrow x = 180° - 68° - 45° = 67$

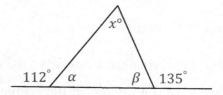

19) Choice D is correct.

$\frac{4,400}{40} = 110$. So 110 square centimeters of wood are needed to make 1 window.

20) Choice B is correct.

Let x be the number of gifts. Therefore, $12.50 per each gift equals $12.5x$.

$20 is for the packing. So, you should add that amount to $12.5x$: $y = 12.5x + 20$

21) Choice B is correct.

First calculate square of -3: $|-5| + 9 \times 2\frac{1}{3} + 9$

Convert mix number to fraction, then multiply to 9: $|-5| + \frac{63}{3} + 9$

Calculate absolute value and add terms: $5 + 21 + 9 = 35$

22) Choice B is correct.

The ratio of men to women is $1: 3$. let x be the total number of men and set a proportion:

$\frac{1}{3} = \frac{x}{24}$

Solve for x: $3x = 24 \rightarrow x = 8$

23) Choice D is correct.

The range is the difference between the lowest and highest values: $22 - 2 = 20$

24) Choice B is correct.

$(2x + 4)°$ and $96°$ are vertical angles. Vertical angles are equal in measure.

Then: $2x + 4$, which simplifies to $96 \rightarrow 2x = 92$, therefore $x = 46°$

25) The answer is 97.6.

Use the area of square formula: $S = a^2$, so $595.36 = a^2$, which gives $a = 24.4$

One side of the square is 24.4 feet.

Use the perimeter of square formula: $P = 4a$, so $P = 4(24.4)$, which results in $P = 97.6$

26) Choice B is correct.

The two-digit numbers must be even, so the only possible two-digit numbers must end in 6, since 6 is the only even digit given in the problem. Since the numbers cannot be repeated, the only possibilities for two-digit even numbers are 76 and 56. Thus, the answer is two possible two-digit numbers.

27) Choices D is correct.

Set proportional ratio to find the number of miles:

$\frac{16\,km}{80\,km} = \frac{10\,mi}{x} \rightarrow 16x = 800 \rightarrow x = 50$ miles

28) The answer is 60.

Jason needs an 75% average to pass for the five exams. Therefore, the sum of 5 exams must be at least $5 \times 75 = 375$.

The sum of 4 exams is: $68 + 72 + 85 + 90 = 315$

The minimum score Jason can earn on his fifth and final test to pass is: $375 - 315 = 60$

29) Choice B is correct.

Let's review all choices:

A. $(-3) + 10 = 5 \rightarrow 7 = 5$. This is not true.

B. $\frac{(-3)+6}{-3} = -1 \rightarrow -1 = -1$. This is true.

C. $\frac{2(-3)+1}{2} = 3 \rightarrow \frac{-5}{2} = 3$. This is not true.

D. $2(-3) - 8 = 3 \rightarrow -14 = 3$. This is not true.

Only choice B is true.

30) Choice C is correct.

$10\% + 60\% = 70\% \rightarrow 100 - 70 = 30$, hence 30% of stamps are dated after 2002.

30% of $1,100 = \frac{30}{100} \times 1,100 = 330$

31) Choice C is correct.

The amount of money that jack earns per hour: $\frac{\$616}{44} = \$14$

The number of additional hours that he needs to work to make enough money is: $\frac{\$826 - \$616}{1.5 \times \$14} = 10$

The total number of hours needed is: $44 + 10 = 54$

32) The answer is 32.40.

$Tax = 0.05 \times \$648 = \$32.40$

33) Choice B is correct.

According to the diagram, $2x + 4 = 10$.

Then subtract 4 from each side, $2x + 4 - 4 = 10 - 4$.

Now, $2x$ must equal to: $6 \rightarrow 2x = 6 \rightarrow x = 3$

34) Choice D is correct.

A. $x = \frac{1}{3} \rightarrow \frac{6}{8} + \frac{1}{3} = \frac{18+8}{24} = \frac{26}{24} \approx 1.08\overline{3} < 2$

B. $x = \frac{3}{5} \rightarrow \frac{6}{8} + \frac{3}{5} = \frac{30+24}{40} = \frac{54}{40} \approx 1.35 < 2$

C. $x = \frac{6}{5} \rightarrow \frac{6}{8} + \frac{6}{5} = \frac{30+48}{40} = \frac{78}{40} \approx 1.95 < 2$

D. $x = \frac{4}{3} \rightarrow \frac{6}{8} + \frac{4}{3} = \frac{18+32}{24} = \frac{50}{24} \approx 2.08\overline{3} > 2$

35) Choice C is correct.

In a triangle, the three interior angles always add up to $180°$.

Therefore: $(3x) + (2x + 2) + 48 = 180 \rightarrow 5x + 50 = 180 \rightarrow 5x = 130 \rightarrow x = 26$

36) Choice B is correct.

Let x be the cost for 35 cans. Write a proportion and solve for x. $\frac{5 \text{ cans}}{\$\,3.40} = \frac{35 \text{ cans}}{x} \Rightarrow$

$x = \frac{3.40 \times 35}{5} \to x = \$23.80$

37) Choice C is correct.

Let n represent a number in the sequence, and let x represent the number that comes just before n.

$n = 4 + 2x \to 84 = 4 + 2x \to 80 = 2x \to x = 40$

38) Choice C is correct.

Compare each score: In Algebra Joe scored 20 out of 25 in Algebra that it means 80% of total

mark. $\frac{20}{25} = \frac{x}{100} \to x = 80$

Joe scored 30 out of 40 in science that it means 75% of total mark. $\frac{30}{40} = \frac{x}{100} \to x = 75$

Joe scored 68 out of 80 in mathematic that it means 85% of total mark. $\frac{68}{80} = \frac{x}{100} \to x = 85$

Therefore, his score in mathematic is higher than his other scores.

39) Choice C is correct.

Use the volume of square pyramid formula:

$V = \frac{1}{3}a^2h \to V = \frac{1}{3}(12\,m)^2 \times 10\,m \to V = 480\,m^3$

40) The answer is $38\frac{2}{5}$.

Robert runs $7\frac{1}{5}$ miles on Saturday and $2(7\frac{1}{5})$ miles on Monday and Wednesday.

Robert wants to run a total of 60 miles this week.

Therefore, subtract $7\frac{1}{5} + 2(7\frac{1}{5})$ from 60.

$60 - \left(7\frac{1}{5} + 2\left(7\frac{1}{5}\right)\right) = 60 - 21\frac{3}{5} = 38\frac{2}{5}$

STAAR Mathematics Practice Test 9
Answers and Explanations

1)   Choice B is correct.

The equation of a line in slope intercept form is: $y = mx + b$. We need to solve for y.

$$2x - y = 12 \rightarrow -y = 12 - 2x \rightarrow y = (12 - 2x) \div (-1) \rightarrow y = 2x - 12$$

The slope of this line is 2. Parallel lines have the same slopes.

2)   Choice C is correct.

First, add $1\frac{1}{6}$ and $3\frac{1}{2}$: $1\frac{1}{6} + 3\frac{1}{2} = 4\frac{1}{6} + \frac{1}{2} = 4\frac{1}{6} + \frac{3}{6} = 4\frac{4}{6} = 4\frac{2}{3}$

Now, subtract $4\frac{2}{3}$ from $6\frac{1}{2}$: $6\frac{1}{2} - 4\frac{2}{3} = \frac{13}{2} - \frac{14}{3} = \frac{39}{6} - \frac{28}{6} = \frac{11}{6} = 1\frac{5}{6}$

Ryan needs another $1\frac{5}{6}$ meters of wood to make the desk.

3)   Choice A is correct.

First, find the sale price. 15% of \$45.00 is \$6.75, so the sale price is \$45.00 − \$6.75 = \$38.25. Next, find the price after Nick's employee discount. 20% × \$38.25 = \$7.65, so, the final price of the shoes is \$38.25 − \$7.65 = \$30.60

4)   Choice B is correct.

Mia has answered $\frac{42}{48}$ of the questions: $\frac{42}{48} \times 100 = 87.5\%$

So, 12.5%$(100 - 87.5 = 12.5)$ of questions have not been answered by Mia.

5)   Choice C is correct.

Add the first 5 numbers. $40 + 45 + 50 + 35 + 55 = 225$. To find the distance traveled in the next 5 hours, multiply the average by the number of hours. $Distance = Average \times Time = 50 \times 5 = 250$. Add both numbers: $250 + 225 = 475$

6)   The answer is 9.50.

Divide \$57.00 by 6: $\frac{57.00}{6} = 9.50$

7) The answer is 200.

Let x be the number of blue marbles. Write the items in the ratio as a fraction:

$\frac{x}{150} = \frac{4}{3} \rightarrow 3x = 600 \rightarrow x = 200$

8) Choice A is correct.

Mode is the value in the list that appears most often. Therefore, the mode is 76.

Write the numbers in order: $30, 35, 38, 54, 60, 76, 76$

The median is the number in the middle. Therefore, the median is 54.

Then subtract 54 from 76: $76 - 54 = 22$

9) Choice C is correct.

x and z are co-linear. y and $5x$ are co-linear. Therefore,

$x + z = y + 5x$, subtract x from both sides, then, $z = y + 4x$

10) Choice A is correct.

Let x be the number of new shoes the team can purchase. Therefore, the team can purchase $120x$.
The team had \$20,000 and spent \$14,000. Now the team can spend on new shoes \$6,000 at most.
Now, write the inequality: $120x + 14,000 \leq 20,000$

11) Choice B is correct.

The model consists of four arrows, each representing 4, hence the corresponding expression is
$4.(4)$.

12) The answer is 7.

Adding 6 to each side of the inequality $4n - 3 \geq 1$ yields the inequality $4n + 3 \geq 7$. Therefore, the least possible value of $4n + 3$ is 7.

13) Choice A is correct.

In the scale model of a building 3 inches represents 45 feet. Therefore, 1.5 feet (18 inches) of the scale model represents 270 feet. To write a proportion and solve:

$$\frac{3\ inches}{45\ feet} = \frac{18\ inches}{x} \rightarrow x = \frac{18 \times 45}{3} = 270\ feet$$

14) Choice B is correct.

The area of a rectangle: $width \times length = 2\ m \times 1\ m$

We know that $1\ m = 100\ cm \rightarrow 200\ cm \times 100\ cm = 20,000\ cm^2 = 100(200)cm^2$

Only choice B is equal to $20,000\ cm^2$

$20,000\ cm^2 = 100(200)\ cm^2$

15) Choice A is correct.

$13 < -3x - 2 < 22 \rightarrow$ Add 2 to all sides. $13 + 2 < -3x - 2 + 2 < 22 + 2$

$\rightarrow 15 < -3x < 24 \rightarrow$ Divide all sides by -3. (Remember that when you divide all sides of an inequality by a negative number, the inequality sign will be swapped. $<$ becomes $>$)

$\frac{15}{-3} > \frac{-3x}{-3} > \frac{24}{-3} \rightarrow -8 < x < -5$

16) Choice C is correct.

Two rectangles are said to be similar if the corresponding sides are in proportion. Only choice C presents similar dimensions to Richard's garden. Simplify the ratio: $6\ m$ by $8\ m \rightarrow 3\ m$ by $4\ m$

17) Choice D is correct.

A reflection is a flip over a line. Notice that each point of the original figure and its image are the same distance away from the line of reflection. Therefore, ABC was reflected across the y −axis.

18) Choice B is correct.

Set a proportion: $\frac{4}{50} = \frac{x}{20} \rightarrow 50x = 4 \times 20 \rightarrow 50x = 80 \rightarrow x = \frac{80}{50} = 1.6\ km$

19) Choice D is correct.

We have, $(3^a)^b = 243 \rightarrow 3^{ab} = 243$.

Then, $243 = 3^5$. So, $3^{ab} = 3^5$. Therefore, ab equals 5.

20) The answer is $\frac{62}{15}$.

$$\frac{1\frac{5}{4} + \frac{1}{3}}{2\frac{1}{2} - \frac{15}{8}} = \frac{\frac{9}{4} + \frac{1}{3}}{\frac{5}{2} - \frac{15}{8}} = \frac{\frac{27 + 4}{12}}{\frac{20 - 15}{8}} = \frac{\frac{31}{12}}{\frac{5}{8}} = \frac{31 \times 8}{12 \times 5} = \frac{31 \times 2}{3 \times 5} = \frac{62}{15}$$

21) Choice B is correct.

Set a proportion: $\dfrac{b}{a} = \dfrac{12}{6} = 2$

The constant of proportionality that relates the number of cakes made, b, to the number of cups of sugar used, a, is 2.

22) Choice C is correct.

Plug in the values of x and y provided in the choices into both equations. Let's start with $2x + 2y = 2$:

A. $(1, 3)$, $2x + 2y = 2 \rightarrow 2 + 6 \neq 2$

B. $(2, 4)$, $2x + y = 2 \rightarrow 4 + 8 \neq 2$

C. $(2, -1)$, $2x + 2y = 2 \rightarrow 4 + (-2) = 2$

D. $(4, -6)$, $2x + 2y = 2 \rightarrow 8 + (-12) \neq 2$

Only choice C is correct.

23) Choice A is correct.

Let y be the total cost and set the equation: $y = 3.5x$

24) Choice C is correct.

The formula for the volume of a a pyramid is $V = \dfrac{1}{3}Bh$.

Since the base of the pyramid is a square, the base area is $6\ cm \times 6\ cm = 36\ cm^2$.

Substitute 36 for B and 15 for h in the formula: $V = \dfrac{1}{3}(36)(15) = 180$

The volume of the square pyramid is $180\ cm^3$.

25) Choice C is correct.

First, find the equation of the line. All lines through the origin are of the form $y = mx$, so the equation is $y = \dfrac{1}{3}x$. Of the given choices, only choice C, $(9, 3)$, satisfies this equation:

$$y = \dfrac{1}{3}x \rightarrow 3 = \dfrac{1}{3}(9) = 3$$

26) Choice C is correct.

Solve for the sum of five numbers:

$Average = \frac{sum\ of\ terms}{number\ of\ terms} \rightarrow 24 = \frac{sum\ of\ 5\ numbers}{5} \rightarrow sum\ of\ 5\ numbers = 24 \times 5 = 120$

The sum of 5 numbers is 120. If a sixth number 42 is added, then the sum of 6 numbers is

$120 + 42 = 162.$

$Average = \frac{sum\ of\ terms}{number\ of\ terms} = \frac{162}{6} = 27$

27) Choice C is correct.

$Probability = \frac{number\ of\ desired\ outcomes}{number\ of\ total\ outcomes} = \frac{5}{2+5+3} = \frac{5}{10} = \frac{1}{2}$

28) Choice B is correct.

$8\ in$ to $6\ m \rightarrow 1\ in : \frac{6}{8}\ m \rightarrow 1\ in : \frac{3}{4}\ m$

29) Choice D is correct.

15% of $160 is $0.15 \times 160 = 24.$

30) Choice C is correct.

The fraction $\frac{2}{5}$ can be written as $\frac{2 \times 20}{5 \times 20} = \frac{40}{100}$, which can be interpreted as forty hundredths, or

0.40.

31) Choice C is correct.

Total amount Ava paid: cost of 6 flower pots + delivery cost = $(6 \times \$9.50) + (\$12.30) =$

$\$57 + \$12.30 = \$69.30$

32) Choice A is correct.

Write a proportion and solve for the missing number.

$\frac{32}{12} = \frac{6}{x} \rightarrow 32x = 6 \times 12 = 72 \rightarrow 32x = 72 \rightarrow x = \frac{72}{32} = 2.25$

33) Choice B is correct.

Use the area of the rectangle formula $(s = a \times b)$.

To find the area of the shaded region subtract the smaller rectangle from bigger rectangle.

$S_1 - S_2 = (10\ ft \times 8\ ft) - (5\ ft \times 8\ ft) \rightarrow S_1 - S_2 = 40\ ft^2$

34) Choice B is correct.

To solve this problem, divide $30.40 by 3.2: $\frac{\$30.40}{3.2} = \$9.50$

35) Choice D is correct.

Five years ago, Amy was three times as old as Mike. Mike is 10 years now.

Therefore, 5 years ago Mike was 5 years. Five years ago, Amy was: $A = 3 \times 5 = 15$

Now Amy is 20 years old: $15 + 5 = 20$

36) Choice D is correct.

Circumference of a circle is $2\pi r \rightarrow A = 2\pi(9) \rightarrow A = 18\pi \rightarrow \pi = \frac{A}{18}$.

37) Choice D is correct.

To solve absolute value equations, write two equations. $x - 10$ could be positive 4, or negative 4.

Therefore, $x - 10 = 4 \rightarrow x = 14$, and $x - 10 = -4 \rightarrow x = 6$.

Find the product of solutions: $6 \times 14 = 84$

38) The answer is $2,250$.

Let x be the original price. Then:

$\$1,912.50 = x - 0.15(x) \rightarrow 1,912.50 = 0.85x \rightarrow x = \frac{1,912.50}{0.85} \rightarrow x = 2,250$

39) Choice C is correct.

The formula for the area of the circle is: $A = \pi r^2$

The area is 36π. Therefore: $A = \pi r^2 \rightarrow 6\pi = \pi r^2$

Divide both sides by π: $36 = r^2 \rightarrow r = 6$

The diameter of a circle is $2 \times$ radius. Then: $Diameter = 2 \times 6 = 12$

40) Choice D is correct.

The spent amount is $60, and the tip is 25%. Then: $tip = 0.25 \times 60 = \$15$

$Final\ price = Selling\ price + tip \rightarrow final\ price = \$60 + \$15 = \$75$

STAAR Mathematics Practice Test 10
Answers and Explanations

1) Choice D is correct.

$y = 5ab + 3b^3$. Plug in the values of a and b to the equation: $a = 2$ and $b = 3$.
$y = 5(2)(3) + 3(3)^3 = 30 + 3(27) = 30 + 81 = 111$

2) Choice D is correct.

Use the simple interest formula: $I = prt \rightarrow I = (\$5,000)(0.06)(3) = \$900$
Nina had to pay back \$5,000 in principle plus \$900 in interest: $\$5,000 + \$900 = \$5,900$

3) Choices B is correct.

When you reflect a point across the $x-$axis, the $x-$coordinate remains the same, but the $y-$coordinate is transformed into its opposite: $B(1, 4) \rightarrow B'(1, -4)$

4) The answer is $\frac{18}{25}$.

So far, Kylie has written $10\% + 18\% = 28\%$ of the entire homework. That means she has $100\% - 28\% = 72\%$ left to write. $72\% = \frac{72}{100} = \frac{18}{25}$

5) Choice C is correct.

$\$9 \times 10 = \$90$.
Petrol use: $10 \times 2 = 20$ liters
Petrol cost: $20 \times \$1 = \$20$
Money earned: $\$90 - \$20 = \$70$

6) Choice A is correct.

The percent of girls taking the drawing class is: $75\% \times \frac{1}{3} = 25\%$

7) Choice C is correct.

$\frac{1\ hour}{18\ coffees} = \frac{x}{1,800} \rightarrow 18 \times x = 1 \times 1,800 \rightarrow 18x = 1,800 \rightarrow x = 100$

It takes 100 hours until she's made 1,800 coffees.

8) Choice D is correct.

Use the volume of the triangular prism formula.

$$V = \frac{1}{2}(length)(width)(height) \rightarrow V = \frac{1}{2} \times 7 \times 5 \times 3 \rightarrow V = 52.5\ m^3$$

9) The answer is 45.

The sum of the measures of angle A and angle B is $180° \rightarrow A + B = 180$. Then:

$$(2x + 6) + 84 = 180 \rightarrow 2x = 180 - 90 \rightarrow 2x = 90 \rightarrow x = 45$$

10) Choice D is correct.

Members paid on the first day: $6 \times (\$5) = \$30$

$28 - 6 = 22$. 22 members paid their dues on the second day.

$22 \times (\$5) = \$110$

$\$110$ was collected in dues on the second day of the week.

11) Choice A is correct.

$8 < x \le 10$, then x cannot be equal to 8.

12) Choice D is correct.

$68 - 25 = 43$

13) Choice C is correct.

1 foot equals to 12 inches or 304.8 millimeters (12×25.4).

Thus, 4 feet equal, $4 \times 304.8 = 1,219.2$ millimeters.

And 8 inches equal, $8 \times 25.4 = 203.2$ millimeters.

Then: $1,219.2\ mm + 203.2\ mm = 1,422.4\ mm$

14) The answer is 80.

Since, E is the midpoint of AB, then the area of all triangles DAE, DEF, CFE, and CBE are equal. Let x be the area of one of the triangles, then: $4x = 160 \rightarrow x = 40$

The area of DEC is 80 ($2x = 2(40) = 80$).

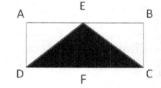

15) Choice D is correct.

$$8a = -4 \rightarrow a = -\frac{1}{2}$$

Let's review all choices:

A. $1 - 6a < 2 \rightarrow 1 - 6(-\frac{1}{2}) < 2 \rightarrow 4 < 2$, This is not true.

B. $-2 + 4a > 2 \rightarrow -2 + 4\left(-\frac{1}{2}\right) > 2 \rightarrow -4 > 2$, This is not true.

C. $-8a + 1 < 0 \rightarrow -8\left(-\frac{1}{2}\right) + 1 < 0 \rightarrow 5 < 0$, This is not true.

D. $-10a - 1 > 3 \rightarrow -10\left(-\frac{1}{2}\right) - 1 > 3 \rightarrow 4 > 3$, This is true!

16) Choice A is correct.

The volume of the cube is less than $64 \ m^3$. Use the formula of volume of cubes.

$volume = (one \ side)^3 \rightarrow 64 > (one \ side)^3$.

Find the cube root of both sides. Then: $4 > the \ one \ side$

The side of the cube is less than 4. Only choice A is less than 4.

17) Choice D is correct.

Write the equation and solve for B: $0.60A = 0.20B$

Divide both sides by 0.20, then: $\frac{0.60}{0.20} A = B$

Therefore, $B = 3A$, and B is 3 times of A or it's 300% of A.

18) The answer is 2.

$\frac{40}{100} = \frac{x}{5} \rightarrow x = \frac{40 \times 5}{100} = 2$

19) Choice D is correct.

$Area = side^2 \rightarrow 121 = side^2 \rightarrow side = 11 \ cm$

$Perimeter = 4 \times side \rightarrow Perimeter = 4 \times 11 \ cm = 44 \ cm$

20) Choice A is correct.

Set up a proportion and solve for x: $\frac{42}{3} = \frac{x}{\frac{1}{7}} \rightarrow 3x = 42 \times \frac{1}{7} \rightarrow 3x = 6 \rightarrow x = 2$ miles

21) Choice B is correct.

$2 \leq x < 6 \rightarrow$ Multiply all sides of the inequality by 3.

Then: $2 \times 3 \leq 3 \times x < 6 \times 3 \rightarrow 6 \leq 3x < 18$

Add 1 to all sides. Then: $\rightarrow 6 + 1 \leq 3x + 1 < 18 + 1 \rightarrow 7 \leq 3x + 1 < 19$

The minimum value of $3x + 1$ is 7.

22) Choice C is correct.

From the choices provided only point $(-2, 3)$ is on the John's route.

23) Choice B is correct.

Set of numbers that are not composite between 1 and 20: $A = \{2, 3, 5, 7, 11, 13, 17, 19\}$

$Probability = \frac{number\ of\ desired\ outcomes}{number\ of\ total\ outcomes} = \frac{8}{20} = \frac{2}{5}$

24) Choice D is correct.

His average speed was: $\frac{1.25}{0.25} = 5$ miles per hour

25) Choice A is correct.

The diameter is, 12. So, the radius is, 6. And $\pi \approx 3.14$.

Use the circumference formula: $C = 2\pi r = 2\pi(6) = 12\pi \approx 37.68 \approx 37.7$

26) Choice B is correct.

Factor the number: $3,375 = 15^3$, $\sqrt[3]{15^3} = 15$, Then: $\sqrt[3]{3,375} = 15$

27) Choice C is correct.

The perimeter of rectangle A is equal to: $2 \times (10 + 6) = 2 \times 16 = 32$

The perimeter of rectangle B is equal to: $2 \times (6 + 4) = 2 \times 10 = 20$

Therefore: $\frac{20}{32} \times 100 = 0.625 \times 100 = 62.5\%$

28) Choice B is correct.

Subtract $\frac{1}{6b}$ and $\frac{1}{b^2}$ from both sides of the equation. Then:

$$\frac{1}{6b^2} + \frac{1}{6b} = \frac{1}{b^2} \rightarrow \frac{1}{6b^2} - \frac{1}{b^2} = -\frac{1}{6b}$$

Multiply both the numerator and denominator of the fraction $\frac{1}{b^2}$ by 6. Then: $\frac{1}{6b^2} - \frac{6}{6b^2} = -\frac{1}{6b}$

Simplify the first side of the equation: $-\frac{5}{6b^2} = -\frac{1}{6b}$

Use cross multiplication method: $30b = 6b^2 \rightarrow 30 = 6b \rightarrow b = 5$

29) Choice B is correct.

Let x and y be the numbers. Then:

$x + y = N$, and $x = 6$. Then $6 + y = N \rightarrow y = N - 6$. Now, $3y = 3(N - 6)$.

30) Choice A is correct.

Let x be the capacity of one tank. Then, $\frac{2}{5}x = 200 \rightarrow x = \frac{200 \times 5}{2} = 500$ Liters

The amount of water in three tanks is equal to: $3 \times 500 = 1,500$ Liters

31) Choice B is correct.

$\frac{1}{5} = 0.2, \frac{5}{3} \approx 1.66, \frac{8}{11} \approx 0.73, \frac{2}{3} \approx 0.66$

$$\frac{5}{3} > \frac{8}{11} > \frac{2}{3} > \frac{1}{5}$$

32) Choice A is correct.

Let the lengths of two sides of the parallelogram be $2x\ cm$ and $3x\ cm$ respectively. Then, its perimeter $= 2(2x + 3x) = 10x$

Therefore, $10x = 40 \rightarrow x = 4$

So, one side $= 2(4) = 8\ cm$ and other side is: $3(4) = 12\ cm$

33) Choice C is correct.

Put $1, 2, 3, 4, , \ldots$ for m in the expression $5m - 1$:

$m = 1 \rightarrow 5(1) - 1 = 4$

$m = 2 \rightarrow 5(2) - 1 = 9$

$m = 3 \rightarrow 5(3) - 1 = 14$

$m = 4 \rightarrow 5(4) - 1 = 19$

Thus, choice C is correct.

34) Choice A is correct.

The range of a set of data is the difference between the highest and lowest values in the set and describes data varied.

35) Choice C is correct.

To find the number of possible outfit combinations, multiply the number of options for each factor:

$3 \times 5 \times 6 = 90$

36) Choice A is correct.

The amount of money for x bookshelf is: $200x$, Then, the total cost of all bookshelves is equal to: $200x + 600$, The total cost, in dollars, per bookshelf is: $\dfrac{Total\ cost}{number\ of\ items} = \dfrac{200x + 600}{x}$

37) Choice D is correct.

Move the square 3 units to the right on the $x-$axis and 5 units down on the $y-$axis. The result is represented by the grid in choice D.

38) Choice C is correct.

Write the ratio of $5a$ to $2b$. $\dfrac{5a}{2b} = \dfrac{1}{10}$. Use cross multiplication and then simplify.

$5a \times 10 = 2b \times 1 \rightarrow 50a = 2b \rightarrow a = \dfrac{2b}{50} = \dfrac{b}{25}$

Now, find the ratio of a to b:

$\dfrac{a}{b} = \dfrac{\frac{b}{25}}{b} \rightarrow \dfrac{b}{25} \div b = \dfrac{b}{25} \times \dfrac{1}{b} = \dfrac{b}{25b} = \dfrac{1}{25}$

39) Choice D is correct.

To answer this question, assign several positive and negative values to x and determine what the value of the expression will be:

x	-2	-1	0	1	2
$2 - x^2$	-2	1	2	1	-2

So, the maximum value of the expression is 2.

40) Choice B is correct.

25% of 120 balls is red. Then: $0.25 \times 120 = 30$

"Effortless Math Education" Publications: Simplifying Mathematics for All

At Effortless Math, our team of dedicated authors is committed to creating and publishing top-notch STAAR Mathematics learning resources, with the goal of making Math more accessible and enjoyable for everyone. We understand the importance of a solid foundation in Mathematics, and we work tirelessly to ensure that our publications provide your students with the most effective learning experience possible, as they prepare for the Grade 7 STAAR Math test.

Our team believes that success in Mathematics can be achieved by anyone, and we are dedicated to empowering students with the tools and resources they need to excel. We take pride in the quality of our publications and are continually working to refine and enhance our materials to better meet the needs of today's learners.

From all of us at Effortless Math, we wish your students the very best of luck and success in their academic journey!

A message for your student: Stay confident, work diligently, and always remember that with the right resources and mindset, you can accomplish anything.

Warmest regards,

The Effortless Math Authors

Author's Final Note

Congratulations on completing this practice book! You've made it to the end – fantastic job!

I would like to extend my heartfelt gratitude for choosing this book to assist you in preparing for your child's STAAR Grade 7 Math test. Amidst a wide array of options, I am truly honored that you opted for this practice book.

It took me years to develop this practice book for the STAAR Grade 7 Math, as I aimed to create a thorough and all-inclusive resource to help students make the best use of their precious time while preparing for the exam.

Drawing from my extensive experience of over a decade in teaching and tutoring math, I have incorporated my personal insights and notes into the development of this book. It is my earnest hope that the information and practice tests provided within these pages will contribute to your child's success in the STAAR Grade 7 Math exam.

If you have any questions, please feel free to contact me at reza@effortlessmath.com, and I will be more than happy to help. Your feedback will enable me to greatly enhance the quality of my future books and make this one even better. Additionally, I acknowledge that there may be a few minor errors in this book. If you come across any, please let me know so that I can rectify them as soon as possible.

If you found value in this book and enjoyed using it, I would love to hear from you. I kindly request that you take a moment to post a review on the book's Amazon page.

I personally read every single review to ensure that my books genuinely benefit students and test takers. By leaving a review, you will help me continue assisting students in their math education.

I wish you and your child all the best in your future endeavors!

Reza Nazari

Math teacher and author

www.EffortlessMath.com

... So Much More Online!

✓ FREE Math lessons

✓ More Math learning books!

✓ Mathematics Worksheets

✓ Online Math Tutors

Need a PDF version of this book?

Please visit EffortlessMath.com

Made in the USA
Las Vegas, NV
16 April 2024

88764977R00149